Biology
A Functional Approach

Study Guide

Biology
A Functional Approach

Study Guide

Julian Brightman B.Sc., Cert.Ed. (Tech), M.A. (Ed.)
Lecturer, Open College of South London at Polytechnic of the South Bank, London

Stella Parker B.Sc., Cert.Ed. (Tech), Ph.D.
Tutor in Continuing Education, City University, London

Nelson

Thomas Nelson and Sons Ltd
Nelson House   Mayfield Road
Walton-on-Thames   Surrey KT12 5PL   UK

51 York Place
Edinburgh   EH1 3JD   UK

Thomas Nelson (Hong Kong) Ltd
Toppan Building 10/F
22A Westlands Road
Quarry Bay   Hong Kong

Distributed in Australia by

Thomas Nelson Australia
480 La Trobe Street
Melbourne   Victoria 3000
and in Sydney, Brisbane, Adelaide and Perth

© J. Brightman and S. Parker 1983

First published by Thomas Nelson and Sons Ltd
1983

Reprinted 1984, 1985, 1987 (twice)

ISBN 0-17-448046-6
NPN 04

All Rights Reserved. This publication is protected in
the United Kingdom by the Copyright Act 1956 and
in other countries by comparable legislation. No part
of it may be reproduced or recorded by any means
without the permission of the publisher. This
prohibition extends (with certain very limited
exceptions) to photocopying and similar processes,
and written permission to make a copy or copies
must therefore be obtained from the publisher in
advance. It is advisable to consult the publisher if
there is any doubt regarding the legality of any
proposed copying.

Printed and bound in Hong Kong

Preface

This book is a Study Guide which is intended to be used in conjunction with *Biology, A Functional Approach* by M.B.V. Roberts.

One purpose of the *Study Guide* is to help the student to make the most effective use of the textbook as an aid to learning. Learning from a book requires specific reading skills, and the *Study Guide* aims to develop these skills by taking the student through a series of structured exercises which are based on the SQ3R (Survey, Question, Read, Recall, Review) process – a process which promotes learning.

Note-making too is an essential aspect of study, and Part 1 of the *Study Guide* describes the techniques of linear and pattern note-making. Again, the exercises in Part 2 of the *Study Guide* are structured in such a way as to promote the development of these note-making skills.

In addition to promoting the acquisition of biological knowledge and the development of effective study techniques, the *Study Guide* aims to provide the student with an understanding of, and ability in, the activities inherent in biological investigation. In Part 3 the student is taken through the processes of observation, measurement, recording, hypothesising, designing experiments and so on by means of exercises which demand active participation. In addition, advice is given on revision and examination techniques.

Thus the *Study Guide*, in conjunction with the textbook, can be used in a variety of ways. For students working largely on their own, on Flexistudy or correspondence courses, for example, the Guide/Textbook combination could form the basis of a course of study. For students attending conventional full- or part-time classes at a school or college, it should prove useful as a private study and revision aid. It could also serve as the basis of an individualised learning approach in the classroom. We hope that it will prove of value to any student using *Biology, A Functional Approach*.

We are indebted to Michael Roberts for his support and encouragement in the production of this *Study Guide* as well as for his comments on the manuscript. We would also like to acknowledge the invaluable advice and assistance provided by Elizabeth Johnston and Donna Evans at Thomas Nelson and Sons Ltd.

J. Brightman
S. Parker
October 1982

Acknowledgements

The authors and publisher wish to thank the following examining bodies for permission to reproduce questions from their past examination papers. The sources are acknowledged in the book by the abbreviations shown in parentheses.

Associated Examining Board (AEB)

Oxford Delegacy of Local Examinations (Oxford)

University of London (London)

Contents

PART 1
Introduction   3

PART 2
Introducing biology   17
Structure and function in cells   21
Tissues, organs and organization   27
Movement in and out of cells   32
The chemicals of life   39
Chemical reactions in cells   44
The release of energy   48
Gaseous exchange in animals   55
Heterotrophic nutrition   60
Autotrophic nutrition   65
Transport in animals   73
Uptake and transport in plants   82
The principles of homeostasis   93
Excretion and osmoregulation   97
Temperature regulation   106
The control of respiratory gases   112
Defence against disease   117
Nervous and hormonal communication   124
Reception of stimuli   135
Effectors   139
Locomotion   144
Behaviour   151
Cell division   158
Reproduction   163
The life cycle   172
Patterns of growth and development   175

The control of growth 182
Mendel and the laws of heredity 188
Chromosomes and genes 193
The nature of the gene 198
Genes and development 204
The organism and its environment 208
Associations between organisms 216
Evolution in evidence 219
The mechanism of evolution 226
Some major steps in evolution 233

PART 3
Techniques of biological investigation 238
Answers to questions 272

PART 1

PART 1

Introduction

1 To the student

This book is a study guide designed to be used in conjunction with the A-level biology textbook *Biology, A Functional Approach* by M.B.V. Roberts (referred to as 'BAFA' from now on). This introduction sets out to explain the purpose of the guide and the way in which it is intended to be used.

The guide is divided into three parts. Part 1 gives some advice on 'study skills' – techniques of reading and note-making which you will need when working with the textbook. It also explains how to use part 2, which forms the bulk of the guide, and which takes the form of a series of questions and exercises geared to the text of BAFA. The questions and exercises are designed to help you to understand and retain the information in each chapter and, if necessary, prepare notes. These questions and exercises are based on a method of reading which has been found to be effective in bringing about learning and they are designed to help you to develop this method. The third part of the guide is intended to help you make the best use of the knowledge you have acquired from your studies. It provides exercises designed to help you develop your skills of observing and recording, presentation and interpretation of biological data, experimental design and examination technique.

The overall aims of the study guide, then, can be summarised as follows. First, to enable students working with BAFA to make the best possible use of it as an aid to learning. Second, to help students to equip themselves with study skills, especially those of reading and note-making. Third, to enable students to communicate effectively their biological knowledge and observations in the form of experimental reports, drawings, graphs, answers to examination questions and so on.

It is not intended that the guide, in conjunction with BAFA, should be seen as a complete course of study in A-level biology.

Biology is essentially a practical subject and any course in biology will involve numerous practical investigations. The companion volume to BAFA, the *Students' Manual*, contains many suggestions for practical work, much of which can only be carried out in a laboratory. Nor is BAFA the only book to which you will need to refer. No one book could possibly encompass such a vast and rapidly growing subject. To become a knowledgeable student of biology you will need to refer to a wide range of books and magazine articles. We do hope, however, that the techniques discussed in the guide will help you to make the most effective use of other sources of information in addition to BAFA.

2 Reading to learn

As a student of biology, you may belong to one of a number of categories. You may be attending full-time or part-time classes at a school or college, you may be doing a correspondence course, or you may simply be working on your own at home. Whatever your cicumstances, any course of study will involve you in work on your own, such as homework, private study and revision. When working on your own, one of your most valuable aids to learning will be your textbook. Making the best use of a textbook is not just a matter of reading it as you would a newspaper or a novel. You use a textbook as an aid to learning and learning involves understanding and retaining information and concepts. It is an active process, requiring concentration and effort, but it can be made easier by adopting effective techniques of learning; by developing study skills.

The most important study skill you will need to develop in order to make the best use of a textbook is that of effective reading. This guide is based on a technique which has been found to work. This technique is called the 'SQ3R'* method, because it suggests that there are five stages in effective reading:

1. Survey (or Scan)
2. Question
3. Read
4. Recall
5. Review

Survey

This stage avoids the danger of 'not seeing the wood for the trees'. By surveying or scanning the chapter or section of the book you are

*See Walker, C. (1974): *Reading Development and Extension*, Ward Lock.

interested in you are able to build up a general idea of what it is about; an overview or conceptual framework. The detailed pieces of information can be fitted into this framework when you come to read in earnest.

The survey stage should not take very long, but it really does help your understanding of the subject matter. Textbooks are usually set out in such a way as to make this stage quite easy. You should turn the pages fairly rapidly, taking in as you do section and subsection headings, words underlined, set in heavy type or italics, and the first and last sentences of each paragraph. You will also find that photographs and diagrams catch your eye, and reading the captions and labels of the illustrations will help you to gain an overview. Many textbooks begin each chapter with an introductory paragraph and conclude with a summary of the main points and it is a good idea to read these at this stage.

Question
As you survey, focus your concentration by mentally posing questions. These questions should occur to you quite readily: Do I understand this? What does this diagram represent? What is the key point of this paragraph? What is the meaning of that unfamiliar word? How do the points being made by successive paragraphs interconnect? and so on. Finding the answer to these, and other more detailed questions becomes your objective for the next stage; reading.

It is a good idea, during the survey and question stages to jot down, in note form, the main points and the questions which occur to you. The study guide provides you with a suggested method for doing this, which is explained on page 9.

Read
Having gained an overview and a set of goals from the surveying and questioning processes, you are now ready to begin reading in earnest. You are now concerned with understanding and absorbing more detailed information. This type of reading requires concentration, and you should find somewhere to work where you can be comfortable but also free from distractions. Read with a view to answering the questions posed in the survey and question stages. Do not try to take in too much at once – read section by section, stopping to **recall** what you have read before going on.

Recall
Having read carefully through a section of text, the next stage is to

try to recall the information in it. This is essentially a self-testing process and can be approached in a number of ways. You can try to write down the main points in note form without looking at the text, or you can use a tape recorder, recalling the main points verbally, or you can ask another person to test you.

This is a very important stage in the learning process, and the study guide contains recall questions designed to help you with it. The answers to these questions can be used to form the basis of a set of notes. If you are not able to answer all of the questions, and it is most unlikely that you will be able to after only one reading, it will be necessary to re-read the relevant section of the text. Repetition is also an important element in the learning process.

Review

In the recall stage, you are testing yourself on material read only a few minutes before. In order to retain information for longer periods, a process of revision is necessary. It has been shown that the amount of information retained by a student after a lesson or reading a book falls away quite rapidly unless the information is reviewed at intervals. Ideally, the first revision should take place within twenty minutes to one hour, but certainly within twenty four hours. This should be followed by a second review after one day, a third after a week, a fourth after a month and a fifth after six months. This may sound rather laborious and it does require discipline and planning, but it is really worthwhile. If you do carry out the first three revisions you should be able to retain about 80 per cent of the information over a long period, but if you do not, you will have forgotten everything within a few weeks. Thus revision is not something to be left until the last moment, but should be done at regular and planned intervals as you go along. This approach makes last minute revision before an examination very much easier.

When reviewing, you do not have to re-read every word of the chapter or section of the book. It is much easier to revise from your own notes prepared during the survey and recall stages of the SQ3R process.

If you are working towards an examination, and most students are, a knowledge of the sort of questions asked in the examination is very important when revising. The study guide provides you with a set of review questions for each chapter which are based on examination questions and are intended to form the basis of a revision programme.

3 Making notes

Making notes is an essential part of study. Note-taking, whether from a lecture, lesson or from a book, involves condensing speech or writing into an abbreviated form, which nevertheless contains the same essential information as the original. This process focuses the concentration and in itself is an aid to learning. The second function of note-taking is to provide yourself with an aid for future study and revision. It is much easier to refer back to, or revise from, notes which you have written and organised yourself in your own personal style than from a textbook. Quite apart from anything else, a set of notes is less weight to carry around than a textbook!

How you obtain your notes will depend upon your circumstances. If you are a student attending classes, you will probably make most of your notes in the classroom and your teacher will be the major source of information. If you are working on a correspondence course or some other form of independent study, you will probably make most of your notes from textbooks and other printed material. The techniques of note-making suggested here can be used in either situation.

Whichever style of note making you eventually develop, there is one pitfall you must try to avoid: do not try to take down every word the teacher says. Similarly, when making notes from written sources, do not copy whole passages. The purpose of note-making is to produce a record of the essential core of information contained in the lesson or printed passage.

Some general advice on note-making

1. Use a loose-leaf file, so that you can remove, add to, or rearrange your notes.

2. Space your notes out on the page. If you leave plenty of room, you can always add further information later.

3. Pick out key words or headings by using colour, underlining and capital letters in a consistent way.

4. Keep your notes as short as possible – use abbreviations, and do not worry about writing perfectly grammatical, complete sentences.

5. **Annotated diagrams** are a good way of presenting information in a condensed form.

Two techniques of note-making
LINEAR OR SKELETON NOTES
Here you write down the key points in a linear sequence, using

Figure A

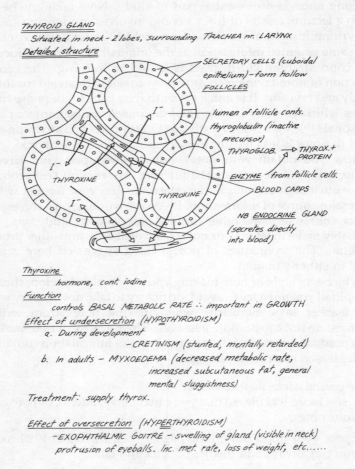

headings, underlining, numbering and colour. Figure A shows part of a set of linear notes prepared after reading through the section in BAFA on the thyroid gland (pages 295–7). If you compare the notes with this passage, you will notice that the information has not only been condensed, it has also been reorganised. This process of re-arranging the information in a way that makes sense to you is an important aid to learning. It also demonstrates the importance of reading the whole of the section before making notes. It is only when you have a clear idea of all the contents that you can decide what is the essential information and what can be left out.

Figure B *Although fewer words have been used, this page actually contains more information than a page of linear notes.*

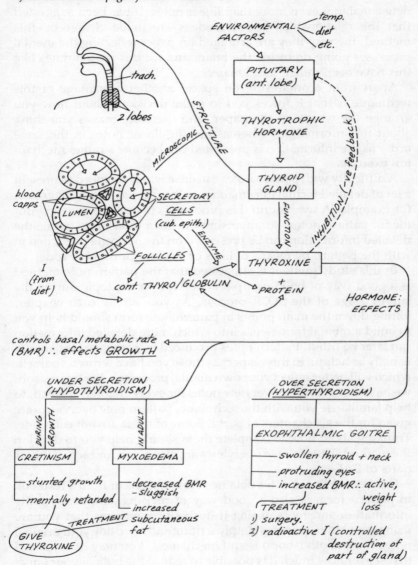

PATTERN NOTES

Figure B shows a set of pattern notes made after reading the same pages in BAFA. In this form of note-making, you write the main topic centrally on the page and then branch out from it. Relationships are indicated by lines, and words are kept to a minimum. As you can see from the example, this method does allow you to condense material even more than linear notes. It has been suggested that the relationships between ideas are made clearer by this method, the way they are arranged on paper reflecting the mental processes going on inside the brain, and for this reason notes like this have been called 'mind maps'.

Apart from economising on space, another advantage of this technique is that it forces you to make decisions about how you arrange the words on the paper, and therefore makes you think about the information, whereas making linear notes in the same order as the information is presented can become a rather mechanical exercise.

A difficulty with pattern notes can arise when you need to present a lot of detailed factual information, in the form of lists or diagrams, for example. A way round this problem is to use one page in your file for pattern notes, summarising the main topic areas, while the detailed information can be presented on the opposite page, tied in with the pattern by means of lines or a colour or number code.

In this study guide it is suggested that the pattern note method is a good way of building up an overview of the topic during the survey stage of the SQ3R process. As you survey each chapter, jotting down the main points in pattern note form should help you to build a mental framework into which more detailed information can later be fitted. Pattern notes produced by someone else are not usually as helpful in this respect as those you have written yourself, which will better reflect your own mental processes. For this reason, we have not provided overview notes for every chapter. Instead, to help familiarise you with the technique, pattern note overviews are given for the six chapters of part I. Some of these are not complete. Thinking about how to complete them should help you to develop the method. Pattern note overviews are also given for each of the six parts of BAFA.

Converting linear notes obtained from lessons or from a textbook to pattern form is also a good way of revising. Condensing the information and reorganising it in this way ensures that you are using your brain, whereas simply writing and rewriting your original notes can be both laborious and mechanical. You may be surprised to discover how much it is possible to reduce the bulk of your linear notes in this way and thus ease the labour of revision.

4 Planning your study

It is generally agreed that most students on a two-year A-level course need to spend about ten hours each week studying each subject. This estimate includes both time spent in the classroom and time spent in private study.

You should aim to organise your private study periods into a regular routine which you should try to stick to. Do not try to do too much at once. Intellectual work is demanding; but it can be made less so if you break your study up into fragments. Study sessions ought to average 45 minutes–1 hour, followed by a break of 5–15 minutes. During the break, get away completely from the work in hand and relax, both mentally and physically. It is a good idea to try to arrange to study in the same place each day, which should be a room as free from distraction as possible, in which all the books and materials you require are to hand.

5 How to use the study guide

The study guide is intended to help you to make the most effective use of BAFA by developing the SQ3R method of reading. For each chapter of BAFA, there is a corresponding section in the study guide giving instructions which take you through the chapter, following the SQ3R process.

The best way to familiarise yourself with the study sequence is to try it out. On the next few pages are given some extracts from the part of the guide dealing with chapter 3 of BAFA; 'Tissues, Organs and Organization', together with notes explaining what you are intended to do at each stage.

Open your copy of BAFA at chapter 3 (page 32), read the instructions overleaf and try to carry them out.

3

Tissues, organs and organization

Survey and question
Survey this chapter. An incomplete set of overview notes is provided in figure 3.1. Attempt to fill in the gaps as you survey. Alternatively, you may wish to produce your own version.*

---

*This instruction only applies to those chapters for which incomplete overview notes are provided (see page 10). For other chapters the instruction is simply 'Survey this chapter and make overview notes in pattern form'.

Survey means reading the introductory paragraph of the chapter and scanning the rest, spending no more than a few minutes on each page.

You should read the first and last sentences of each paragraph and take in the items that capture your attention, such as:
Section headings, e.g. Classification of tissues
                      Epithelium: a simple animal tissue
and so on.

Words or sentences printed in heavy type, e.g. **tissue, organs, epithelial, connective, blood,** etc.

Photographs and diagrams and their captions.

Read
CLASSIFICATION OF TISSUES
EPITHELIUM: A SIMPLE ANIMAL TISSUE

The next step, the first of the Rs, is to read. Not the whole chapter at once, but section by section, reading carefully. You are now concerned with understanding and absorbing more detailed information, and this kind of reading requires a high degree of concentration, so it is best not to try to rush or to take in too much at once. *However*, if you do feel confident that you have understood and retained the subject matter easily, there is no reason why you should not read several sections together.

> *Recall*
>
> Make a copy of the outline classification of epithelial tissues in Figure C and complete it by filling in the gaps. Add simple diagrams illustrating the main types.

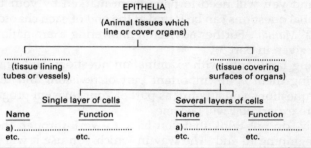

The aims of the recall questions are:
1 to help you to check that you have understood *and retained* the information in the section.
2 to help you to learn the information by organising it in a different way.
3 to provide you with a summary of the main points which you can incorporate into your notes.

You may be able to answer the questions without referring back to BAFA, and you can use them as a self-test when revising. However, if you are studying the topic for the first time, it is unlikely that you will have retained all the details after one reading, and you will need to refer back to the section in order to find the answer. Re-reading parts of the section in order to extract the information you require should help you to learn.

The read and recall stages are repeated for each section of the chapter.

> *Review*
>
> 1 Give an illustrated account of (a) a parenchyma cell, (b) a sieve tube cell, (c) a nematoblast and (d) a glandular epithelial cell. [5, 5, 5, 5]
> 2 What is meant by (a) a cell and (b) a tissue? [4, 4]
>   Describe the structure of xylem and bone, explaining in each case how structure is related to function. [12]
> 3 Write an essay on the variety of animal cells. [20]

The final stage in the SQ3R process is review. For each chapter of BAFA, the study guide provides a selection of Review questions, which should serve to focus your revision; marks for each question are given in brackets. They are mostly based on examination questions, and you will need to have them marked by your teacher. Additional questions can be found at the end of each chapter in the *Students' Manual*. Further advice on answering examination questions is given in part 3.

Gaining familiarity with examination questions and practice in answering them is an important part of revision, and the more review questions you attempt as part of your revision programme, the more expert you will become.

Finally, do not think that you have to work through the guide from beginning to end. The way in which you use it will depend upon the way in which you are using BAFA. If you are studying a topic for the first time and BAFA is your only source of information, you will probably benefit from working through the whole of the study guide for that chapter. However, if you have studied the topic before in class and you wish only to check a specific point, you do not need to work through the guide for the whole chapter, and may not wish to go through the survey and question process if you are confident that you already have an overview. Similarly, you do not *have* to use the answers to the recall questions as the basis for your notes if you have already made notes in class. It is up to you to develop the style of study which is best suited to your personality and needs. While no-one else can do this for you we do hope that you will find the suggestions in this book helpful.

PART 2

1
Introducing biology

Survey and question

Survey this chapter. A completed set of overview notes in pattern form is provided in figure 1.1. Relate these notes to the contents of the chapter as you survey. You may copy these notes or produce your own if you prefer.

Overview notes

Figure 1.1

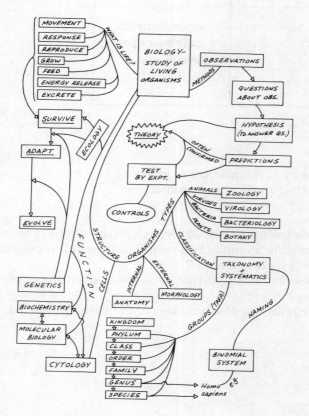

Read
THE DIFFERENT BRANCHES OF BIOLOGY

Recall
1 a) What is biology?
   b) List and describe each of the different branches of biology.

Read
WHAT IS LIFE?

Recall
1 Make a table which summarises the seven characteristics of living things. You may find table 1.1 a useful guide.

Table 1.1 *The characteristics of living things*

Characteristic	
1 Movement etc.	Carried out by all organisms, or parts of organisms. etc.

Read
METHODS OF INVESTIGATION IN BIOLOGY
CONDUCTING BIOLOGICAL EXPERIMENTS

Recall
1 a) Analyse the described research on diphtheria in terms of the scientific method. You may find it useful to organise your analysis as in table 1.2.
   b) Describe a control that could be used in the experimental stage.
   c) What are the disadvantages of the scientific method?

Table 1.2

Stage in the scientific method	Application to problem
1 Observation etc.	Mouse injected with diphtheria dies. Bacteria remain isolated at injection site. etc.

Read

ORGANISMS

Recall

1 a) Arrange the taxonomic groupings of organisms in a hierarchy. Part of such a hierarchy is given in figure 1.2.
   b) Classify a human according to this system.

Figure 1.2

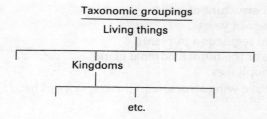

2 a) Who devised the binomial system of classification?
   b) What are the advantages of this system?

Read

SOME BASIC CONCEPTS

Recall

1 Write *brief* notes on (a) survival,
   (b) adaptation,
   (c) the relationship between structure and function,
   (d) evolution.

In each case give a named example to illustrate your point.

Review

1. Design a controlled experiment to investigate *either*
   a) the effect of temperature on the rate of growth of bean seedlings, *or*
   b) whether houseflies have colour vision. *[20]*

2. To what extent can (a) a candle flame and (b) a virus be considered to be living? *[10, 10]*

3. To which branch of biology do each of the following scientific papers belong?

   Courtship in sticklebacks.
   Some laboratory techniques in bacteriology.
   Feeding, assimilation and digestion in locusts.
   A study of grasslands.
   The regulation of body fluids in animals.
   A new species of *Ranunculus* from Patagonia.
   A comparative study of the mouthparts of three species of wasps.
   Heredity in pea plants.
   The microstructure of cells.
   The biology of yeasts.
   Hormonal regulation in plants.
   Structure of the human adrenal gland.
   T_4 bacteriophages.
   Comparative wing structure in three species of bat. *[1 mark each]*

2

Structure and function in cells

This is the first chapter in part I 'Organization in cells and organisms'. The overview notes in figure 2.1 show how the topics covered by the chapters in this part of the book are related.

Overview notes for part I: Organization in cells and organisms

Figure 2.1

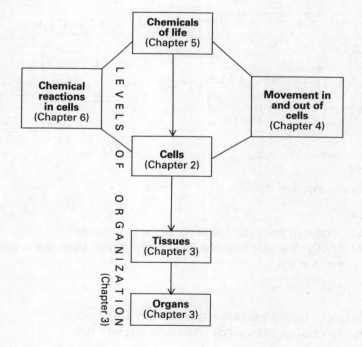

Survey and question

Survey this chapter. An incomplete set of overview notes is provided in figure 2.2 Attempt to fill in the gaps as you survey. Alternatively, you may prefer to produce your own version.

Overview notes

Figure 2.2

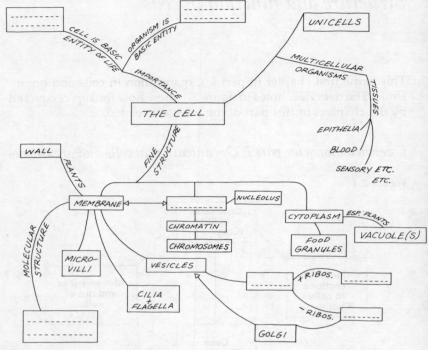

Read

THE CELL THEORY

Recall

1. a) State two important aspects of the cell theory.
   b) Briefly describe how the cell theory differs from the organismal theory.

Read

AN ANIMAL CELL AS SEEN WITH THE LIGHT MICROSCOPE
A PLANT CELL AS SEEN WITH THE LIGHT MICROSCOPE

Recall

1. Make an annotated diagram of a generalised cell showing those structures which can be seen using the light microscope. In your annotations, indicate clearly those structures which can be found only in plant cells.

Read
OTHER KINDS OF MICROSCOPE

Recall
1. a) Explain why higher magnifications can be obtained with the electron microscope than with the light microscope.
   b) Name four techniques which can be used to make cell structures more visible with the light microscope.
   c) Describe the disadvantages of using the electron microscope to examine cell structure, and explain how they may be, to some extent, overcome.

Read
FINE STRUCTURE OF THE CELL (page 16-25)

Recall
1. Describe how cell organelles (a) can be isolated from cells and (b) their functions investigated.
2. Make brief notes on the structure and function of the following ultrastructural components of cells:
   a) limiting membranes,
   b) secretory membrane systems,
   c) organelles concerned with carbohydrate metabolism,
   d) organelles which digest worn-out cell components,
   e) organelles concerned with cell movement and locomotion.

You may find it useful to tabulate your notes as shown in table 2.1:

Table 2.1

Structure	Function
a) *Limiting membranes* e.g. Plasma membrane A unit membrane composed of phospholipid and protein etc.	Regulates passage of materials in and out of cell. etc.

Read

THE MOLECULAR STRUCTURE OF THE PLASMA MEMBRANE (pages 26–8)

Recall

1. **a)** Describe the observations and evidence used by Danielli and Davson to formulate their hypothesis of membrane structure.
   b) Describe three lines of evidence which support their predictions about the structure of the plasma membrane.

2. **a)** By means of annotated diagrams, describe the differences in structure between the Danielli–Davson model and the fluid–mosaic model of the cell membrane.
   b) Describe two lines of evidence which led Singer and Nicholson to put forward the fluid–mosaic model.

Read

THE DIVERSITY OF CELLS

Recall

1. **a)** Account for the differences in the structure of the cell types in a multicellular organism.
   b) Copy and complete table 2.2 which lists some of the cell types and their functions in a multicellular animal.

   Table 2.2 *The diversity of cells*

Type of cell	Function
Epithelial etc.	Lines the surface of organs and cavities. etc.

2. How do you account for (a) the similarities, and (b) the differences between animals and plants?

Review

1. Write concise notes on each of the following: (a) mitochondria, (b) lysosomes and (c) endoplasmic recticulum. (d) How has electron microscopy advanced our knowledge and understanding of the structure of cells? [5, 5, 5, 5]

2 Describe the techniques involved in investigating the structure of cell organelles. For any one organelle, describe how its structure can be related to its function. [8, 12]

3 Write an essay on cell membranes. [20]

4 Describe the roles of the following in biosynthesis:
   a) Golgi apparatus, [5]
   b) endoplasmic reticulum, [7]
   c) chloroplasts. [8]

5 Figure 2.3 is a drawing of part of an electronmicrograph of root tip cells (magnification 15 000×). Examine the drawing, and answer the following questions, referring to any structure you name by giving its letter on the drawing.

   Figure 2.3

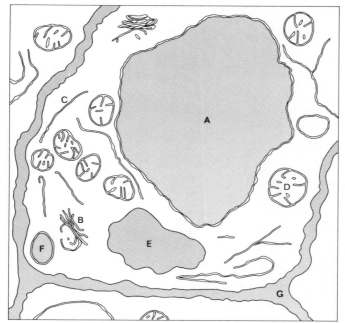

   a) Name *two* features which indicate the material is of plant origin. [2]
   b) Name *two* features which indicate that the cells are meristematic. [2]
   c) Name a structural feature which at least *three* organelles in the drawing have in common. Identify the organelles selected. [4]

d) How are the functions of the organelles **A**, **B** and **C** related? [3]
e) Relate the structure of the organelles **D** to their function and suggest why so many seem to be present in this particular cell. [5]
f) State *two* problems associated with the use of the electron microscope as an instrument to gain information about the ultrastructure of living cells. [4]

3

Tissues, organs and organization

Survey and question
Survey this chapter. An incomplete set of overview notes is provided in figure 3.1. Attempt to fill in the gaps as you survey. Alternatively, you may wish to produce your own version.

Overview notes

Figure 3.1

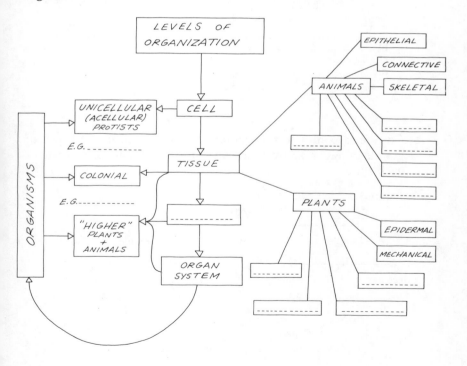

Read

CLASSIFICATION OF TISSUES
EPITHELIUM: A SIMPLE ANIMAL TISSUE

Recall

1 Make a copy of the outline classification of epithelial tissues in figure 3.2 and complete it by filling in the gaps. Add simple diagrams illustrating the main types.

Figure 3.2

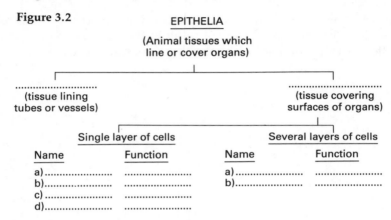

2 Copy and complete the classification of the different types of glands in figure 3.3 by giving an example and simple diagram of each type.

Figure 3.3

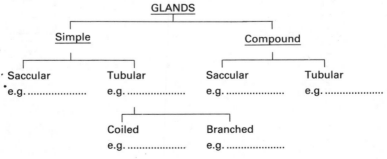

Read

CONNECTIVE TISSUE
SKELETAL TISSUE

Recall

1 a) Make a copy of figure 3.4 which summarises the constituents of areolar tissue. Complete it by filling in the gaps.

Figure 3.4

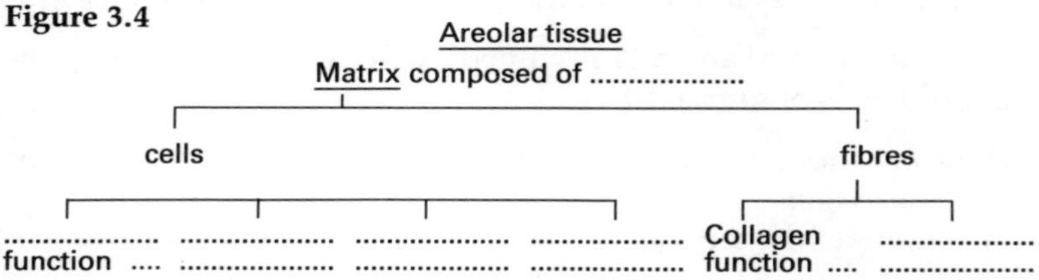

```
                         Areolar tissue
                  Matrix composed of ....................
        ┌─────────────────┴─────────────────┐
       cells                               fibres
  ┌─────┬─────┬─────┐                   ┌────┴────┐
  ..........  ..........  ..........  ..........   Collagen   ..........
  function ....  ..........  ..........  ..........   function ....  ..........
```

b) Construct similar diagrams for:
  (i) collagen tissue,
  (ii) elastic tissue,
  (iii) cartilage,
  (iv) bone.

Read

OTHER ANIMAL TISSUES
PLANT TISSUES
THE DIFFERENCES BETWEEN ANIMAL AND PLANT TISSUES

Recall

1 Copy and complete table 3.1 to compare and contrast animal and plant tissues.

Table 3.1

	Animals		Plants	
Tissue	Name	Function	Name	Function
1 Transporting tissue	Blood Red cells	carry O_2	Vascular tissue	
	White cells	combat disease	Phloem	carries food
etc.			Xylem	carries water

Read

ORGANS AND ORGAN SYSTEMS
LEVELS OF ORGANIZATION
COLONIAL ORGANIZATION
THE ADVANTAGES OF THE MULTICELLULAR STATE

Recall

1 Using named examples, define (a) organ and (b) organ system.

2 Using plants and animals as named examples, explain the meaning of levels of organisation.

3 a) From memory, make a large, fully labelled diagram of *Paramecium*.
   b) Annotate the diagram to describe the functions of the following structures; macronucleus, micronucleus, cilia, contractile vacuole, trichocysts, oral vestibule, food vacuole, cytoproct.

4 a) From memory, make a large, fully labelled diagram of the body wall of *Hydra*.
   b) Annotate the diagram to describe the functions of the following types of cell: musculo-epithelial, glandular, sensory, nematoblasts, nerve, interstitial.

5 a) From memory, make a large, fully labelled diagram of the body wall of a sponge.
   b) Annotate the diagram to describe the functions of the following types of cell: mesenchyme, collar, amoebocytes, hollow pore, epithelial.

6 Explain each of the following:
   a) *Hydra* is on the tissue level of organisation.
   b) A sponge is on the colonial level of organisation.
   c) A higher plant is organised on the tissue level.
   d) *Volvox* is on the colonial level of organisation.

7 List (a) the advantages, and (b) the disadvantages of the multicellular state.

Review

1 Give an illustrated account of (a) a parenchyma cell, (b) a sieve tube cell, (c) a nematoblast and (d) a glandular epithelial cell. [5, 5, 5, 5]

2 What is meant by (a) a cell and (b) a tissue? [4, 4]
   Describe the structure of xylem and bone, explaining in each case how structure is related to function. [12]

3 Write an essay on the variety of animal cells. [20]
4 a) Describe the structure and function of epithelial cells in a mammal. [10]
   b) Compare and contrast the structure and function of the outer epidermis in *Hydra* and a mammal. [10]

4

Movement in and out of cells

Survey and question

Survey this chapter. An incomplete set of overview notes is provided in figure 4.1. Attempt to fill in the gaps as you survey. Alternatively, you may wish to produce your own version.

Overview notes

Figure 4.1

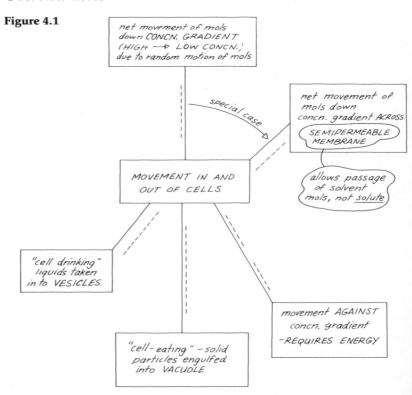

Read

DIFFUSION

DIFFUSION AND THE STRUCTURE OF ORGANISMS

Recall

1 When a lump of sugar is dropped into a cup of tea it dissolves. Describe the physical processes which cause it to dissolve.

2 Name three metabolic processes which depend on diffusion.

3 Giving two named examples of each, describe how concentration gradients (a) arise in living organisms, and (b) can be increased in living organisms.

4 Calculate the surface area : volume ratio of the three cubes in figure 4.2.

Figure 4.2

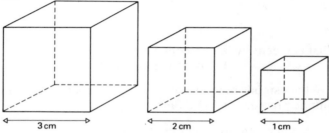

5 What is the biological significance of surface area : volume ratio?

6 In most animals, adaptations have evolved which result in diffusion-dependent processes being more efficient. Describe in note form the nature of these adaptations in the following examples: *Hydra*, flatworms, lungs, gills, the insect tracheal system.

Read

OSMOSIS

CELLS AS OSMOMETERS

Recall

1 a) In what way is osmosis a special case of diffusion?
   b) Describe the conditions necessary for osmosis to occur in cells.

c) Which of the following is capable of developing the greater osmotic pressure: (i) a weak sugar solution, (ii) a strong sugar solution. Explain your answer.

2 In figure 4.2 in BAFA, explain (a) why osmotic pressure develops in the thistle funnel, (b) how this pressure can be detected, (c) how this pressure could be prevented from building up, (d) the meaning of osmotic potential and (e) the meaning of osmometer.

3 Explain what would happen if (a) red blood cells and (b) an *Amoeba* living in fresh water were each put into the following: (i) an isotonic solution, (ii) a hypotonic solution and (iii) a hypertonic solution.

Read

WATER POTENTIAL

Recall

1 What is meant by (a) osmotic pressure and
                 (b) osmotic potential?

2 Give the osmotic potential of (a) pure water and (b) a cell with an osmotic pressure of 600 kN/m^2.

3 a) What is meant by water potential?
   b) How could you lower the water potential of a solution?

Read

OSMOSIS AND PLANT CELLS

Recall

1 'Plant cells generally have a solute concentration which is markedly higher than that of their surroundings.' What is meant by their surroundings?

2 Explain how each of the following develops in plant cells, (a) full turgor and (b) full plasmolysis.

Read

PLANT WATER RELATIONS
WILTING

Recall
1 Write an equation which summarises the water relations of a plant cell.
2 a) Describe the appearance of a plant cell for which
   (i) WP is zero,
   (ii) WP is equal and opposite to OP.
   b) For each of these cells, what is the value of Ψ?
3 Describe how you could measure each of the following for plant cells (a) WP and (b) OP.
4 How does wilting differ from plasmolysis?

Read
ACTIVE TRANSPORT

Recall
1 Using three pieces of evidence to support your explanation, describe how active transport differs from diffusion.
2 a) Summarise, in note form, two hypotheses which could account for active transport.
   b) List five properties of cell membranes necessary for active transport to occur.
3 What evidence is there to suggest that, during active transport of ions into cells (a) some carriers are able to transport different ions and (b) some carriers transport only one kind of ion?

Read
PHAGOCYTOSIS
PINOCYTOSIS

Recall
1 Construct a table which compares and contrasts phagocytosis and pinocytosis.

Review
1 a) Explain the meaning of:
   (i) osmosis,
   (ii) osmotic pressure (osmotic potential),
   (iii) suction pressure (or diffusion pressure deficit or water potential),
   (iv) wall pressure (or pressure potential). [8]

b) Give an equation which shows the relationship between (ii), (iii) and (iv). [2]
c) Describe methods by which you could determine the following for plant tissue:
   (i) the osmotic pressure (osmotic potential).
   (ii) the suction pressure (diffusion pressure deficit, water potential). [10]

2 a) A dandelion stalk was split longitudinally and divided into three pieces of length 3 cm, as shown in figure 4.3.

Figure 4.3

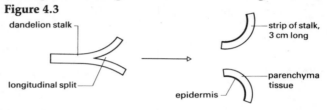

One piece was placed in distilled water and after five minutes it appeared as in figure 4.4.

Figure 4.4

(i) Explain the reason for the change in the shape of the piece of stalk.

Another piece of stalk was placed in 2 per cent sucrose solution and after five minutes it appeared as in figure 4.5.

Figure 4.5

(ii) What does this result indicate about the relationship between 2 per cent sucrose solution and the cell sap of parenchyma cells?

A third piece of stalk was placed in 10 per cent sucrose solution.

(iii) Draw a diagram to show its appearance after five minutes immersion, and explain why the stalk becomes this shape. [9]

b) Table 4.1 gives values for the initial osmotic potential (OP) and wall pressure (WP) of plant cells which were then bathed in solutions of given osmotic potential. For each example given calculate (i) the initial DPD of the cell and (ii) the net direction of water flow. [6]

Table 4.1

OP of bathing solution	Plant cells at start	
	Initial OP	Initial WP
− 300 kN/m²	− 1010 kN/m²	600 kN/m²
− 500 kN/m²	− 1120 kN/m²	750 kN/m²
0 kN/m²	− 550 kN/m²	0 kN/m²

c) Blood samples were taken from a mammal and an equal volume added to three tubes: **A**, **B** and **C**.

Tube **A** contained 15 g salt/litre.
Tube **B** contained 10 g salt/litre.
Tube **C** contained 5 g salt/litre.

The contents of the tubes were gently mixed, allowed to stand for one hour and then centrifuged.
 (i) Explain why the salt solution in tube **C** was pink, whilst it remained colourless in tubes **A** and **B**.
 (ii) The red blood cells (deposited by centrifugation) in tube **B** appeared normal, whilst those in tubes **A** and **C** were abnormal. Make diagrams to show the appearance of red blood cells in tubes **A** and **C**. Give explanations for their appearance. [5]

3 a) Define diffusion. [3]
 b) State two factors that affect the rate of diffusion in living organisms. [2]
 c) Calculate the surface area : volume ratio of cube **A**. [2]

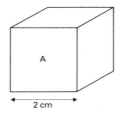

 d) Draw diagrams showing how the surface area : volume ratio of cube **A** could be increased, without altering its volume. [5]

e) Explain the significance of the following:
   (i) Tapeworms have flattened bodies.
   (ii) Plants inhabiting dry environments have stomata sunken in pits.
   (iii) Mammalian red blood corpuscles are flattened, biconcave discs.
   (iv) The inner lining of the mammalian small intestine consists of finger-like projections, known as villi. [8]

4 Discuss the role of: (a) diffusion and (b) turgor in the lives of animals and plants. [10, 10]

5 Using a named example in each case, describe the importance of the following in the lives of animals:
   a) active transport,
   b) phagocytosis,
   c) pinocytosis. [6, 7, 7]

6 Discuss the different methods by which materials move into and out of cells. [20]

5
The chemicals of life

Survey and question
Survey this chapter. An incomplete set of overview notes is provided in figure 5.1. Attempt to fill in the gaps as you survey. Alternatively, you may wish to provide your own version.

Overview notes

Figure 5.1

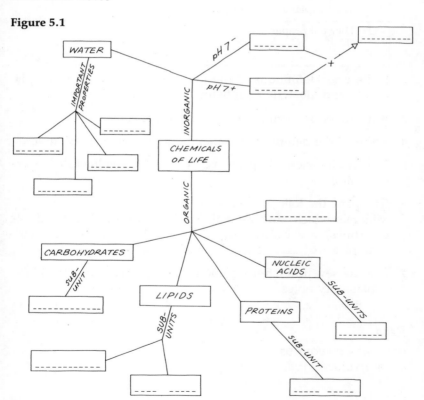

Read

WATER
ACIDS AND BASES
SALTS (pages 58–62)

Recall

1 Construct a table describing *four* important properties of water and their biological significance along the lines suggested in table 5.1.

   Table 5.1

Property	Biological significance
1 Solvent properties (brief explanatory notes)	
2 Thermal properties etc.	

2 Write brief definitions of the following terms: acid, base, pH, buffer, and illustrate each definition with an example.

3 Why are buffers important to living organisms?

4 Write a brief definition of mineral salt, and give an example.

5 List *five* biologically important cations and *five* biologically important anions.

6 Complete the following statement:
   Minerals which are essential, but are only needed in minute quantities, are called _____ _____. Two examples are _____ and _____.

7 List the seven functions of salts in living organisms, with brief explanatory notes.

Read

ORGANIC COMPOUNDS
CARBOHYDRATES (pages 63–9)

Recall

1 What is (a) the valency of carbon, and
          (b) its three-dimensional shape?

2 The carbon skeleton of organic molecules may have different shapes. List three shapes that are commonly found.
3 List three elements other than carbon which are commonly found in organic compounds.
4 Make a large copy of table 5.2, and complete it to summarise the main properties of carbohydrates. Alternatively, present the information in pattern note form (see 'Proteins' page 42).

Table 5.2

	Monosaccharides	Disaccharides	Polysaccharides
General formula			
Structure of molecule (Brief description or diagram of e.g.)			
Examples (formulae, where found, etc.)			
Properties			
Functions			
etc.			

Read

LIPIDS (pages 69–72)

Recall

1 By means of *either* a table *or* pattern notes make a summary of the structure, occurrence, properties and functions of lipids.
An outline pattern for proteins is given in figure 5.2 which may be used as a guide.

Read

PROTEINS (pages 73–9)

Recall

1 Make a summary of the structure, occurrence, properties and functions of proteins in pattern note form. A suggested outline is given in figure 5.2. Alternatively, tabulate the information as you did for carbohydrates.

Figure 5.2

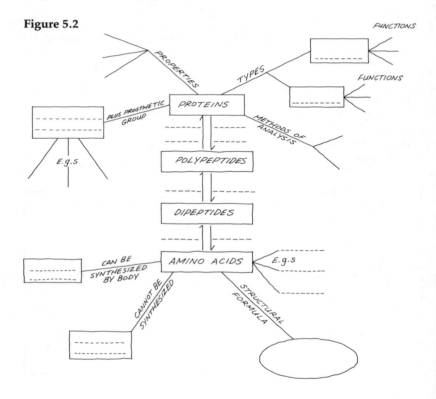

Read

VITAMINS

Recall

1 Construct a table giving the following information:
   a) the names of the main vitamins,
   b) the kinds of foods they are found in,
   c) the effects of deficiency.

Review

1. **a)** Distinguish between:
   - (i) fats and oils,
   - (ii) starch and cellulose,
   - (iii) haemoglobin and myoglobin,
   - (iv) sucrose and maltose. [8]

   b) Describe, giving full experimental details, how you would carry out simple tests on an unknown substance to detect the presence of carbohydrates, lipids and proteins. [12]

2. Discuss the importance of proteins to animals and plants. [20]

3. Discuss the properties of water in terms of its importance as a medium for life. [20]

4. Describe the importance of the following in the metabolism of a normal mammal:
   - **a)** mineral salts,
   - **b)** steroids,
   - **c)** vitamins. [6, 6, 8]

5. **a)** Distinguish chemically between carbohydrates and fats. [5]

   b) Describe how chemical structure is related to the functions of each of the following:
   - (i) glucose,
   - (ii) glycogen,
   - (iii) cellulose,
   - (iv) triglycerides,
   - (v) phospholipids. [15]

6

Chemical reactions in cells

Survey and question

Survey this chapter. An incomplete set of overview notes is provided in figure 6.1. Attempt to fill in the gaps as you survey. Alternatively, you may wish to produce your own version.

Overview notes

Figure 6.1

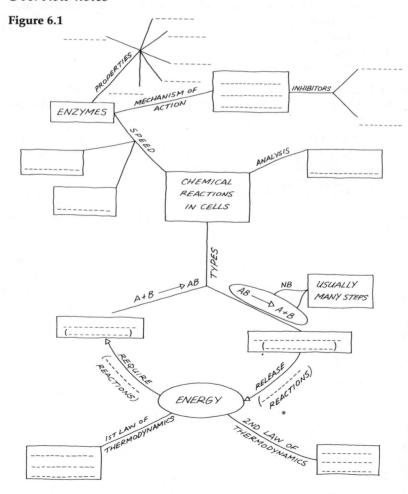

Chemical reactions in cells

Read
ANALYSING METABOLIC PATHWAYS
TYPES OF CHEMICAL REACTIONS IN CELLS

Recall

1. Define (a) metabolism, (b) metabolite and (c) metabolic pathway.
2. Describe techniques used to (a) identify intermediates in metabolic pathways and (b) trace the fates of various elements in metabolism.
3. Using one named example for each, describe (a) a synthesis reaction and (b) a breakdown reaction.
4. a) Describe three examples of anabolic, endergonic pathways in *both* animals and plants.
   b) Suggest why plants and bacteria have greater synthetic powers than animals.

Read
ENERGY
ENERGY CONVERSIONS
ACTIVATION ENERGY
THE SPEED OF BIOLOGICAL REACTIONS

Recall

1. a) Write an equation summarising the most important catabolic, exergonic reaction in cells.
   b) For this reaction, summarise diagrammatically the relationships between activation energy, reaction energy, potential energy and kinetic energy.
   c) Describe the main uses for energy transferred from catabolic, exergonic reactions in cells.
2. a) Name the chemical process by which plants can transfer energy from sunlight.
   b) Why are animals absolutely dependent upon plants?
3. State (a) the first law of thermodynamics, (b) the second law of thermodynamics and (c) explain how metabolic reactions can result in the complete recycling of matter, but not of energy.
4. List three factors that can affect the speed of a biochemical reaction.
5. a) Define catalyst.
   b) Give another name for biological catalysts.

Read

ENZYMES
NAMING AND CLASSIFYING ENZYMES
THE PROPERTIES OF ENZYMES

Recall

1 a) Describe how enzymes were first discovered and (b) state two functions of enzymes in cells.

2 b) State two criteria for classifying enzymes.
   b) Describe how enzymes are named.

3 For each of the following metabolic processes, name the groups of enzymes responsible for catalysing the reactions in: (a) respiration and (b) digestion.

4 Make a table which compares and contrasts the seven properties of enzymes with those of inorganic catalysts.

Read

HOW ENZYMES WORK
ENZYME INHIBITORS
ALLOSTERIC ENZYMES
COFACTORS AND PROSTHETIC GROUPS

Recall

1 a) Describe a hypothesis to explain enzyme action.
   b) List four lines of evidence which support this hypothesis.

2 a) Explain enzyme inhibition and describe the properties of an enzyme inhibitor.
   b) What role do enzyme inhibitors play in metabolism?

3 Using named examples, describe the differences between (a) competitive inhibition, (b) non-competitive inhibition and (c) allosteric inhibition.

4 Using named examples, describe the differences between (a) cofactor, (b) coenzyme and (c) prosthetic group.

Review

1 a) Compare and contrast the properties of enzymes and inorganic catalysts. [5]
   b) Using named examples, explain the meaning of denaturation, specificity and inhibition of enzymes. [15]

2 Give an account of the role of enzymes in the regulation of cellular activities. [20]

3 a) Give an account of the factors that affect the rate of enzyme reactions. [8]
b) Describe how you could investigate the effect of two of these factors on the rates of named enzyme-controlled reactions. [12]

4 a) What is an enzyme? [4]
b) Describe how enzymes are classified. [6]
c) Briefly discuss the role of coenzymes and cofactors in metabolism. [10]

5 You have been provided with three solutions A, B and C.
One contains starch, one contains a disaccharide sugar and one salivary amylase.
Describe, giving full experimental details, how you would:
a) determine which solution is which. [15]
b) investigate one named property of the enzyme. [5]

7

The release of energy

This is the first chapter in part II 'The maintenance of life'. The overview notes in figure 7.1 show how the topics covered by the chapters in this part of the book are related.

Overview notes for part II: The maintenance of life

Figure 7.1

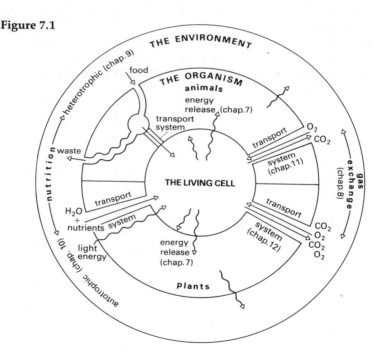

Survey and question
Survey this chapter and make overview notes in pattern form.

Read
THE GENERAL NATURE OF RESPIRATION

Recall

1 a) Write an equation which summarises respiration.
   b) Why is this equation misleading?

2 How could you demonstrate that respiration requires O_2 and produces CO_2 in (a) a small mammal and (b) a plant?

Read

THE COMPOSITION OF INSPIRED AND EXPIRED AIR
STUDYING THE CONSUMPTION OF OXYGEN IN HUMANS
MEASURING THE OXYGEN CONSUMPTION OF A SMALL ORGANISM

Recall

1 a) Describe, in note form, a procedure for collecting and analysing samples of expired air in humans.
   b) In note form, describe and explain how the composition of expired air varies under different conditions.

2 For humans, (a) how can the rate of oxygen consumption be calculated? (b) How is it possible to relate oxygen consumption to energy expenditure?

3 a) By means of a simple annotated diagram, describe a method of determining direct oxygen uptake and carbon dioxide production in a small mammal.
   b) How does this method differ from the Haldane method?

Read

RESPIRATORY QUOTIENT
MEASURING THE ENERGY OUTPUT OF THE BODY
THE METABOLIC RATE IN DIFFERENT CIRCUMSTANCES
FOOD AND ENERGY
MAN'S ENERGY REQUIREMENTS

Recall

1 a) At STP, equal volumes of gas contain equal numbers of molecules. Using the equation for the oxidation of glucose (respiration) calculate the ratio of the volume of carbon dioxide produced to that of oxygen consumed. This is the respiratory quotient (RQ).
   b) Explain why RQs are of limited value in determining the energy output of the body.

2 a) Define basal metabolic rate (BMR).
   b) In note form, explain why BMR may vary in an individual.
3 a) Describe, in note form how you would determine a person's daily energy budget.
   b) List the possible sources of error in your method.
4 Table 7.1 summarises the main classes of food and the ways in which they are used by the body. Make a copy and complete it.

Table 7.1 *The main classes of food and their uses in the body*

Class of food	Energy value (in decreasing order)	Use of food by body
Fat etc.	38.5 kJ/g etc.	Energy store etc.

Read

THE VITAL LINK: ATP

Recall

1 a) In respiration, the energy released by the oxidation of sugar is coupled to the synthesis of ATP. Summarise this diagrammatically.
   b) List the pieces of evidence which suggest that ATP supplies energy in biological systems.

Read

ENERGY FROM SUGAR

THE HYDROGEN CARRIER SYSTEM

Recall

1 Make a copy of table 7.2 and complete it.

Table 7.2 *A summary of the exergonic oxidation reactions of the hydrogen and electron carrier systems*

Type of exergonic oxidative reaction	Enzymes involved	Carrier system involved
Removal of hydrogen atoms (dehydrogenation) etc.	Dehydrogenase etc.	NAD etc.

2 Name an additional source of energy to those in the table above.

3 What happens to the energy released as a result of these reactions?

Read
THE PATHWAY BY WHICH SUGAR IS BROKEN DOWN
GLYCOLYSIS
THE CITRIC ACID CYCLE

Recall
1 Describe, in note form, two experimental techniques for investigating the biochemical pathways of respiration.

2 Summarise the respiration of glucose in a table such as table 7.3 overleaf.

Read
RESPIRATION WITHOUT OXYGEN (pages 106–10)

Recall
1 What is meant by (a) complete anaerobe, and (b) partial anaerobe?

2 List some named examples of complete and partial anaerobes.

3 Write equations to compare the respiration of glucose in (a) anaerobic conditions, and (b) aerobic conditions.

4 a) By means of chemical equations show how pyruvic acid is produced during respiration in plants and animals.
   b) Indicate the fate of the pyruvic acid in the above examples.

Read
ENERGY FROM FAT
ENERGY FROM PROTEIN
THE ROLE OF ATP
THE CONTROL OF SUGAR BREAKDOWN

Recall
1 List the circumstances under which (a) fats and (b) proteins may be utilised for energy release.

2 Describe briefly how fats and proteins are converted into compounds which then enter the pathways of carbohydrate metabolism.

Table 7.3 The respiration of glucose

	Reactants	Products (state number of C atoms)	Type of reaction	Site of reaction	Number of ATP molecules synthesised
Glycolysis	Glucose + ATP	Phosphorylated sugar (6C)	Phosphorylation	Cytoplasm	0
	etc.	etc.	etc.	etc.	etc.
Kreb's cycle					

3 What evidence led Barbara Banks to suggest that ATP performs a homeostatic role in the cell?
4 Show how Barbara Banks' views on the role of ATP in cell metabolism differ from the traditional view.

Review

1 a) List the main differences between aerobic and anaerobic respiration. [6]
   b) Describe how you could show that:
      (i) oxygen is needed for respiration,
      (ii) yeast cells can respire anaerobically. [7, 7]

2 a) Explain the essential features of each of these processes in aerobic respiration:
      (i) glycolysis, [5]
      (ii) Kreb's cycle, [5]
      (iii) electron-transfer system. [5]
   b) Outline briefly the differences between the anaerobic and aerobic respiration of glucose. [5]

3 a) Explain the meaning of phosphorylation. [2]
   b) Briefly account for the different quantities of ATP product in aerobic and anaerobic respiration. [6]
   c) Describe the role of ATP and ADP in cell metabolism. [12]

4 The apparatus shown in figure 7.2 can be used to measure the rate of respiration in small animals.

Figure 7.2

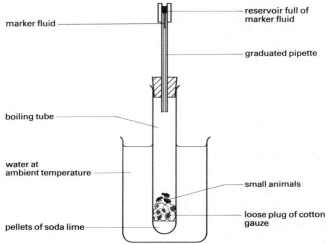

a) Explain how this apparatus works. [6]
b) List the factors that would influence the rate of oxygen uptake. [4]
c) Describe, giving experimental details, how you could use the apparatus to measure oxygen uptake by a named organism. [6]
d) List any sources of error which might arise. [4]

5 Respiration is a characteristic of living things. Discuss. [20]

6 a) Distinguish between:
   (i) anabolism and catabolism,
   (ii) NAD and ATP,
   (iii) hydrogen transfer and electron transfer. [12]
b) Using named examples, describe when you expect to find the following:
   (i) RQ = 1
   (ii) RQ > 1
   (iii) RQ < 1 [8]

8
Gaseous exchange in animals

Survey and question
Survey this chapter and make overview notes in pattern form.

Read
SPECIALIZED RESPIRATORY SURFACES

Recall
1 By means of an annotated diagram, describe how gas exchange occurs in a named protozoan.
2 Using named examples, make annotated diagrams to show how the respiratory surface area has been increased in some animals.
3 What is the biological significance of this increase?

Read
GASEOUS EXCHANGE IN MAN
STRUCTURE OF THE LUNGS

Recall
1 Make an annotated diagram of a named mammal which shows:
   a) the structure of the respiratory tract and associated organs.
   b) the details of the blood supply at the respiratory epithelium.
   c) the routes taken by oxygen and carbon dioxide during gas exchange.
2 a) By means of annotated diagrams, describe the ventilation mechanism in a human.
   b) Indicate on your diagram the pressure changes that occur in the lungs and pleural cavities.

Read
THE BREATHING CYCLE
EXCHANGES ACROSS THE ALVEOLAR SURFACE

Recall

1. a) Assuming your tidal volume to be 0.5 litres, calculate your ventilation rate.
   b) Make a diagram to show the different lung volumes in a human.
   c) Describe in detail how you could measure
      (i) the inspiratory reserve volume,
      (ii) the expiratory reserve volume,
      (iii) the inspiratory capacity,
      (iv) the vital capacity of the lungs.
   d) How do each of these change under conditions of exercise?
   e) Define residual volume.

2. Calculate the percentage of inspired air (a) which remains in the dead space of the respiratory system, and (b) which enters those parts of the lungs where gas exchange is possible.

3. In note form, describe and explain the following:
   a) the difference in composition between inspired air and expired air.
   b) the difference in the partial pressures of oxygen and carbon dioxide in blood flowing (i) to an alveolus and (ii) from an alveolus.
   c) why the composition of alveolar air remains relatively unchanged.
   d) why blood leaving the lungs is not as fully oxygenated as it might be.

Read

GASEOUS EXCHANGE IN FISHES
STRUCTURE OF THE DOGFISH'S GILLS
COUNTER FLOW AND PARALLEL FLOW
GASEOUS EXCHANGE IN TELEOST FISHES
VENTILATION OF THE GILLS

Recall

1. By means of annotated diagrams, compare and contrast,
   a) the ventilation mechanisms in a dogfish and a bony fish.
   b) the flow of the respiratory medium over the respiratory surface of a dogfish and a bony fish, indicating why the latter system is more efficient.

Gaseous exchange in animals 57

Read

GASEOUS EXCHANGE IN INSECTS

Recall

1 Make an annotated diagram of the respiratory system in an insect to show (a) a tracheal trunk and tracheoles, (b) how gas exchange occurs under resting conditions and (c) how gas exchange takes place under conditions of muscular activity.

Read

THE NERVOUS CONTROL OF BREATHING

Recall

1 a) Copy figure 8.1 into your notes.

Figure 8.1 *The nervous control of breathing in a mammal*

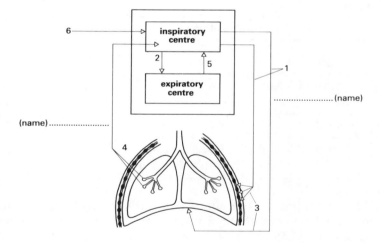

b) Below is a list of processes which occur during the nervous control of breathing. These events are arranged in a random order. Select from the list the processes which correspond to the numbers 1–6 on figure 8.1 and write them in next to the number.
    (i) Inspiratory centre in hindbrain sends impulses to expiratory centre.
    (ii) Inspiratory centre sends nerve impulses to diaphragm and intercostal muscles.

(iii) Impulses relayed from stretch receptors in the lungs to the inspiratory centre.
(iv) Diaphragm and intercostal muscles contract, causing inspiration.
(v) Inspiration is inhibited; expiration (a largely passive process) commences.
(vi) Rising CO_2 levels in the blood stimulate the inspiratory centre.
c) Explain the effect on breathing of cutting the vagus nerve.

Review

1. a) List the main characteristics of a surface used for gaseous exchange. [4]
   b) Describe gas exchange in an earthworm, an insect and a flowering plant. [16]
2. Some properties of water and air are given in table 8.1.

Table 8.1

	Water	Air
Oxygen content	about 1%	20%
Oxygen diffusion rate	low	high
Viscosity	× 100 that of air	low
Density	Specific gravity × 1000 that of air	very low

What problems do the above data pose when terrestrial and aquatic animals ventilate their respiratory surfaces? Illustrate how these problems have been overcome by referring to the mechanisms of ventilation used by a variety of *named* organisms. (AEB)

3. a) Distinguish between:
   (i) ventilation,
   (ii) gas exchange,
   (iii) cell (internal) respiration. [3]
   b) Describe the mechanism of ventilation in a mammal. [10]
   c) Explain how ventilation is controlled in a mammal. [7]
4. Compare and contrast the mechanism of gas exchange in an insect and a named fish. [20]

5 a) Explain, giving full experimental details, how you would investigate gas exchange in:
   (i) a small mammal,
   (ii) an aquatic, submerged green plant (e.g. *Elodea*). [16]
   b) Explain the effects on each of a change in the ambient temperature. [4]

9
Heterotrophic nutrition

Survey and question
Survey this chapter and make overview notes in pattern form.

Read
GENERAL CONSIDERATIONS
PHYSICAL AND CHEMICAL DIGESTION
HOW DIGESTIVE ENZYMES WORK
ABSORPTION AND ASSIMILATION
THE SITE OF DIGESTION

Recall
1. If you have not already done so as part of your overview notes,
    a) write a concise definition of heterotrophic nutrition.
    b) list the three types of heterotrophic nutrition, giving a brief explanation of each.
2. Copy and complete figure 9.1.
3. Name the main kind of chemical reaction involved in digestion, and explain briefly what happens in this kind of reaction.
4. The process of digestion generally requires several groups of enzymes acting consecutively. List, in order, the four groups of enzymes necessary for the complete digestion of proteins and explain their action.
5. List three ways in which the mammalian intestine is well adapted to absorb the products of digestion.
6. Give an example of each of the following:
    a) an animal which carries out intracellular digestion,
    b) an animal with both extracellular and intracellular digestion,
    c) an animal in which digestion is almost entirely extracellular.

Figure 9.1 *Processes involved in heterotrophic nutrition*

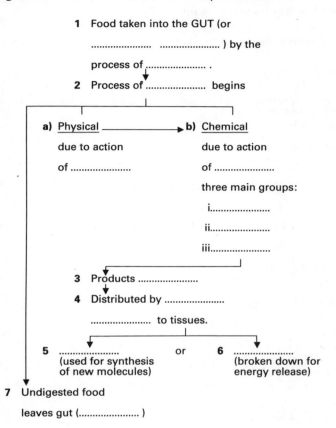

Read
FEEDING AND DIGESTION IN MAN
IN THE BUCCAL CAVITY
DOWN TO THE STOMACH
INTO THE DUODENUM
ABSORPTION

Recall
1 From memory, make a large, fully labelled diagram to show the main features of the human alimentary canal and associated organs. Check your diagram against figure 9.2 on page 127 of BAFA.

2 Summarise, in note form, the information on pages 127–33. A suggested outline is given in figure 9.2.

Figure 9.2 *Digestion and absorption in man*

A. Digestion in buccal cavity
   i. Food broken down by action of teeth
      (Include diagram of section through human molar if you feel you need to refresh your memory of tooth structure)
   ii. Moistened by ....

   contains: (a) ....

   (b) ....

   (c) .... (Enzyme, begins digestion

   of .... to ....)

   iii. .... formed, swallowed and passed to stomach via

   .... by ....

B. Digestion in Stomach
   Gastric Juice

   contains: (a) Enzymes

   i. .... (name and functions)

   ii. .... (name and functions)

   (b) .... (name and functions)

   (c) .... (name and functions)

C. Digestion in the Duodenum
   (names, sources and functions of enzymes and bile)

D. Absorption
   etc.

Read

NUTRITION IN OTHER HETEROTROPHS
HERBIVORES
CARNIVOROUS ANIMALS
CARNIVOROUS PLANTS
LIQUID FEEDERS
MICROPHAGOUS FEEDERS

Recall

1. Classify heterotrophs into four categories based on their diet.
2. a) Herbivores face a particular problem associated with the nature of their food. What is it?
   b) By means of simple annotated diagrams, show how the mouth-parts of a mammal, an insect and a mollusc are adapted to a herbivorous diet.
   c) Give two examples of symbiotic relationships involving cellulase-producing micro-organisms.
3. a) Carnivores face a particular problem associated with the nature of their food. What is it?
   b) List three examples of ways in which animals are adapted to a carnivorous diet.
   c) List three ways in which carnivores may deal with their food after capturing it, giving an example of each.
4. a) Name three British and two non-British carnivorous plants.
   b) The habitats of insectivorous plants have one feature in common. What is it?
5. a) Copy and complete the summary in figure 9.3.

Figure 9.3

```
                       LIQUID FEEDERS
         ┌─────────────────┴─────────────────┐
    ............                        ............
E.g. (1).................           E.g. (1).................
    (2).................                (2).................
```

b) Identify a problem associated with blood as a food source when compared with plant juices, and explain how it is overcome.

6. Giving named examples, describe in note form the main features of microphagous feeders.

Review

1. a) List the main processes involved in nutrition in a mammal. [4]
   b) State the role of the accessory organs in digestion in a mammal. [4]
   c) Briefly describe the roles of the following in mammalian digestion:
   buccal cavity, stomach, small intestine, large intestine. [12]

2   Discuss the range of feeding methods used by animals. [20]

3   Experiments on the rate of absorption of sugars by a rat intestine produced the data in table 9.1.

Table 9.1

Sugars	Relative rates of absorption taking normal glucose uptake as 100	
	By living intestine	By intestine poisoned with cyanide
Hexose sugars: glucose	100	30
galactose	106	35
Pentose sugars: xylose	32	32
arabinose	30	31

a) Suggest a reason for the difference between the rates of absorption of hexose and pentose sugars in the living intestine. [2]
b) Describe the mechanisms by which hexose sugars are absorbed by living intestines. [5]
c) Suggest an explanation for the effect of cyanide on the mechanism of hexose absorption. [2]
d) Suggest *two* experiments to verify your explanation in (c). [6]
e) What are the advantages to the individual of having hexose sugars absorbed in this way? [1]
f) In an intact mammal, absorption of fatty acids is drastically curtailed by any clinical condition which leads to a reduction in bile salt secretion or release. Explain why this is so. [4]

(AEB)

4   a) Make a fully labelled diagram of the digestive system of a named mammal. [8]
    b) Describe how the structure of this digestive system is related to the mammal's mode of nutrition. [12]

5   a) Give a full account of digestion in a named herbivore and a named carnivore. [16]
    b) Explain how the herbivore obtains protein and the carnivore obtains carbohydrates. [4]

10

Autotrophic nutrition

Survey and question
Survey this chapter and make overview notes in pattern form.

Read
DIFFERENT TYPES OF AUTOTROPHIC NUTRITION
THE IMPORTANCE OF PHOTOSYNTHESIS
THE RAW MATERIALS OF PHOTOSYNTHESIS
THE PRODUCTS OF PHOTOSYNTHESIS

Recall
1 Write brief definitions of:
   a) autotrophic (holophytic) nutrition,
   b) photosynthesis,
   c) chemosynthesis.
2 Copy and complete the diagram of the carbon cycle in figure 10.1.

Figure 10.1

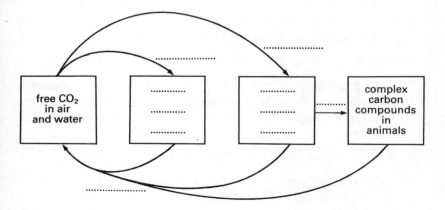

3 List the three essential raw materials of photosynthesis, and indicate in note form why they are essential.

4 Describe how you would demonstrate that (a) starch, and (b) oxygen are produced as a result of photosynthesis.

Read

THE CONDITIONS REQUIRED FOR PHOTOSYNTHESIS
CARBON DIOXIDE
WATER
LIGHT
TEMPERATURE
CHLOROPHYLL
INTERACTION OF FACTORS CONTROLLING PHOTOSYNTHESIS
PHOTOSYNTHESIS AND THE ENVIRONMENT

Recall

1 List the five factors necessary for photosynthesis.

2 When demonstrating the necessity of these factors experimentally, what has to be done to the plant before commencing the experiment?

3 Describe how you would demonstrate that carbon dioxide is necessary for photosynthesis. Mention in your account (a) controls to be used, (b) precautions to be taken and (c) possible sources of error.

4 Why is it difficult to demonstrate simply that water is necessary for photosynthesis?

5 Describe how you would demonstrate that (a) light, and (b) chlorophyll are essential for photosynthesis.

6 Why are leaves green?

7 Make a careful copy of figure 10.7 (page 143 in BAFA). Indicate the colours of the various wavelengths on the x-axis. Annotate the graph to explain the separation of the curves at the 480 nm position.

8 Write a concise statement of the law of limiting factors.

9 Make a careful copy of figure 10.10 (page 145 in BAFA). Annotate the graphs to indicate which factor is limiting in each set of circumstances.

10 a) List three circumstances under which light is likely to be the factor limiting photosynthesis.
   b) In what circumstances is CO_2 availability likely to be the factor limiting photosynthesis?
   c) When is temperature likely to limit photosynthesis?

11 Define (a) compensation point, and (b) compensation period.

Read

THE SITE OF PHOTOSYNTHESIS
THE CHLOROPLAST PIGMENTS
STRUCTURE OF THE CHLOROPLAST
STRUCTURE OF THE LEAF
THE LEAF AS AN ORGAN OF PHOTOSYNTHESIS

Recall

1 Draw a diagram, similar to figure 10.12 in BAFA, to illustrate Engelmann's experiment. Annotate it to explain the method, results and conclusions.

2 List the five chlorophyll pigments that may be separated by paper chromatography. Indicate their colour and the wavelengths of light they absorb as outlined in table 10.1.

Table 10.1

Pigment	Colour	Wavelengths absorbed
1 Chlorophyll a etc.	blue-green etc.	red, blue/violet etc.

3 Copy and complete the following notes.

Chlorophyll belongs to a group of compounds called .... , which also includes ....
Both molecules contain a metal at the centre. In chlorophyll, this is .... , while in .... it is ....

4 Using figures 10.16, 10.17 and 10.18 in BAFA as a guide, describe the structure of chloroplasts by means of annotated diagrams. The annotations should include information about the functions and size of the various parts.

5 Using figures 10.20 and 10.21 in BAFA, illustrate leaf structure by a large annotated low power plan of a transverse section through a leaf. Do not draw individual cells but indicate the different tissue regions in your diagram. Include the leaf midrib. Describe the structure of the mesophyll by means of a separate high power drawing of not more than six cells.

Read

THE CHEMISTRY OF PHOTOSYNTHESIS
PHOTOSYNTHESIS AS A TWO-STAGE PROCESS
THE LIGHT STAGE

Recall

1 List two pieces of evidence that photosynthesis is a two-stage process.

2 By means of notes or a diagram, summarise the main features of the light and dark stages. Indicate the necessary prerequisites for each stage to occur and the end products of each stage.

3 In note form, describe an experiment to demonstrate that the oxygen produced in photosynthesis is derived from water rather than carbon dioxide.

4 Briefly describe the two functions of the light stage.

5 Construct a table along the lines shown in table 10.2, to compare and contrast oxidative phosphorylation and photophosphorylation. (Refer back to BAFA, page 103 to revise oxidative phosphorylation if necessary.)

Table 10.2

	Oxidative phosphorylation	Photo-phosphorylation
1 Raw materials 2 End products 3 Carriers 4 What is carried? 5 Source of electrons 6 Number of pathways 7 Cyclic/non-cyclic		

6 List the differences between cyclic and non-cyclic photo-phosphorylation in terms of raw materials and end products.

7 Memorise and reproduce in your notes the left-hand side of figure 10.26 from BAFA which summarises the reactions of the light stage of photosynthesis.

Read
THE DARK STAGE
THE EXACT SITES OF LIGHT AND DARK STAGES
C_3 AND C_4 PLANTS
PHOTORESPIRATION

Recall

1 Copy and complete the following summary of the dark stage:

1 Essentially the dark stage involves the reduction of .... to form ....
2 This is an .... process.
3 The energy is supplied by the .... of the .... produced in the light stage, and also by ...., which now carries high energy ....
4 The .... also provides the .... for reducing the ....

2 Using figure 10.28 and the information given on page 156 in BAFA, make an annotated diagram of Calvin's apparatus. List the stages in the method.

3 Memorise and reproduce in your notes the right-hand side of figure 10.26 in BAFA which summarises the dark stage.

4 On your summary diagrams of the light and dark stages, make a note of the exact sites of these reactions.

5 The equation in figure 10.2 shows the carbon dioxide fixation stage of the Calvin cycle in C_3 plants.

Figure 10.1

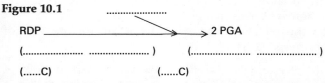

RDP ⎯⎯⎯⎯⎯⎯⎯⎯→ 2 PGA
(...................... ....)   (...................... ....)
(......C)   (......C)

Complete it and construct an equivalent equation for C_4 plants.

6 List two reasons why C_4 plants are more efficient at fixing carbon dioxide than C_3 plants.

Read

AUTOTROPHIC BACTERIA
PHOTOSYNTHETIC BACTERIA
CHEMOSYNTHETIC BACTERIA
FIXATION OF NITROGEN

Recall

1 List three ways in which photosynthesis in purple sulphur bacteria differs from photosynthesis in higher green plants.

2 In terms of their mode of nutrition, what is the essential difference between the purple and green sulphur bacteria and the colourless sulphur bacteria?

3 What are the roles in the nitrogen cycle of the following:
   1 soil saprophytes,
   2 *Nitrobacter*,
   3 symbiotic bacteria in root nodules,
   4 *Nitrosomonas*,
   5 higher plants,
   6 *Nitrococcus*,
   7 animals?

Use your answers to construct a summary diagram of the nitrogen cycle.

Review

1 a) List the functions of ATP in the metabolism of cells and explain how these functions are carried out. [8]
  b) Give a brief outline of ATP production in the cells of a green plant in the light. [8]
  c) State briefly how the plant carries out ATP production in the dark. [4]

2 a) Give a full account of the influence of external factors on the rate of photosynthesis in a green plant. [10]
  b) Describe the effect of *two* external factors on the rate of photosynthesis. [10]

3 a) The plant physiologist F.F. Blackman developed the idea of limiting factors. The effect of one such limiting factor is illustrated in figure 10.3.

Figure 10.3

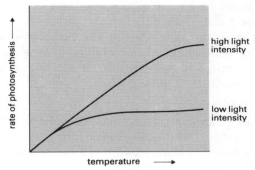

(i) With reference to figure 10.3 explain what is meant by the term *limiting factor*. [2]
(ii) State *one* other limiting factor involved in photosynthesis. [1]
(iii) Explain how you would set up a controlled experiment to investigate this limiting factor. [5]

b) Seven algal cultures were grown under constant conditions in the laboratory. After two days six of the cultures were transferred to a lake and suspended at various depths. The seventh culture was subcultured into fresh media on the eighth day. Figure 10.4 shows the results of the experiment.

Figure 10.4

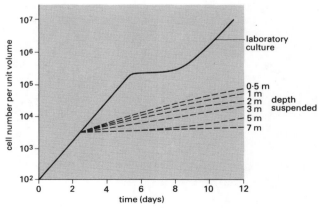

(after J. W. G. Lund)

(i) Suggest why the population growth rate of the laboratory culture slowed down after 5 days. [2]
(ii) Describe how your hypothesis could be tested. [2]

(iii) State *two* factors which are most likely to account for the changes in growth rate when the cultures were transferred to the lake. [2]

(iv) From your knowledge of the biochemistry of photosynthesis explain precisely how *one* of the factors causing the change in growth rate in the lake is involved in photosynthesis. [6]

(AEB)

4   a) By means of annotated diagrams, describe the structure of a leaf and a tuber in a potato plant. [10]

b) Describe how carbohydrates are synthesised, translocated and stored in a potato plant. [10]

5   Write an essay on C_4 plants. [20]

6   Discuss how biologists could alleviate famine and malnutrition by improving the present production of foodstuffs and developing new sources of food. [20]

11

Transport in animals

Survey and question
Survey this chapter and make overview notes in pattern form.

Read
BLOOD
RED BLOOD CELLS

Recall
1   a) Describe some of the functions of transport systems in animals.
    b) What is meant by a circulatory system?
2   Copy and complete figure 11.1 (page 74) which summarises the constituents of mammalian blood.
3   Suggest reasons why red blood cells (a) have no nucleus and (b) are biconcave, flattened discs.
4   a) By means of a simple diagram describe the chemical nature of haemoglobin (Hb) (Read 'The haemoglobin molecule as a carrier', page 167).
    b) Annotate the diagram to explain why Hb has such a high affinity for oxygen.

Read
CARRIAGE OF OXYGEN
MYOGLOBIN AND OTHER BLOOD PIGMENTS
THE HAEMOGLOBIN MOLECULE AS A CARRIER

Recall
1   a) Describe the nature of the union between Hb and oxygen.
    b) Discuss the biological advantages of this type of union.
    c) How could you demonstrate experimentally that Hb has a high affinity for oxygen?

Figure 11.1 *The constituents of mammalian blood*

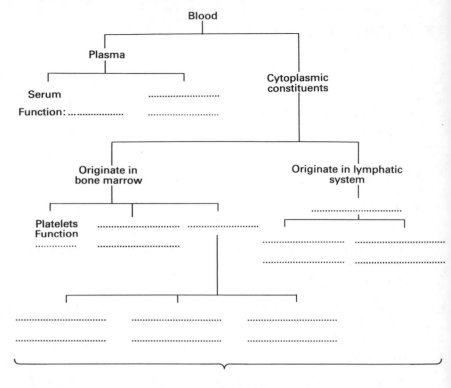

Leucocytes (white blood cells)

2 In which regions of the body would you expect to find oxygen at (a) low partial pressures (oxygen tensions) and (b) high partial pressures?

3 a) Examine figure 11.4 in BAFA. Assume the broken line to be an oxygen dissociation curve for a hypothetical solution **A**. Using the information in this figure, copy and fill in table 11.1 to compare the percentage saturation with oxygen of blood and solution **A**.

   b) Compare blood and solution **A** in terms of their efficiency in combining with oxygen within the range of partial pressures found in the body, and explain why Hb is a more efficient respiratory pigment than solution **A**.

Table 11.1 *A comparison of the percentage saturation with oxygen of blood and solution **A** at a range of partial pressures.*

Partial pressures (kN/m^2)	2.0	4.0	6.0	8.0	10.0
Blood					
Solution A					

4 a) Under what conditions in the body would the oxygen dissociation curve be shifted to the right?
   b) What effect would this shift have on the affinity between oxygen and Hb?

5 Give explanations for the following observations.
   a) Myoglobin is found abundantly in the muscles of seals and whales.
   b) Lugworms are able to respire aerobically in oxygen-deficient mud.
   c) The oxygen dissociation curve for foetal Hb is to the left of the adult Hb curve.
   d) Carbon monoxide is a fatal poison.

6 Copy and complete table 11.2.

Table 11.2 *A summary of the main respiratory pigments found in the animal kingdom.*

Pigment	Prosthetic group	Where found
Hb etc.	Iron etc.	Vertebrates and ... etc.

Read

THE CARRIAGE OF CARBON DIOXIDE

Recall

1 Describe how CO_2 is transported (a) in the plasma and (b) in the red blood cells.

2 Define (a) the Bohr effect and (b) the chloride shift.

Read

THE MAMMALIAN CIRCULATION

Recall

1 Copy figure 11.2 and complete the labelling.

Figure 11.2

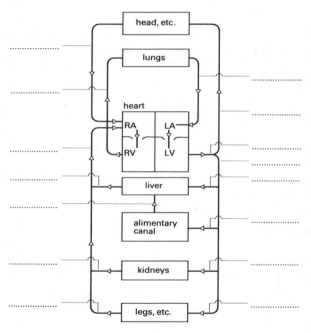

Read

THE HEART

Recall

1 Copy figure 11.3 and complete the labelling.
2 Using this diagram as a model, construct two annotated diagrams to show the events of the cardiac cycle.

Figure 11.3

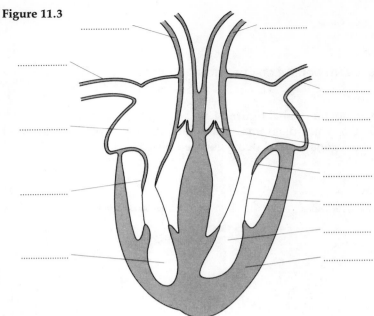

Read

THE BEATING OF THE HEART
INNERVATION OF THE HEART

Recall

1 What evidence is there to suggest (a) that cardiac muscle is myogenic and (b) that the SAN acts as a pacemaker?
2 Draw a diagram showing the innervation of the heart. Annotate your diagram to show how these nerves control the SAN.

Read

BLOOD FLOW THROUGH ARTERIES AND VEINS
THE CAPILLARIES

Recall

1 Copy and complete table 11.3.

Table 11.3

	Structure of wall	Rate of blood flow	Direction of blood flow	Function of vessels
Arteries				
Veins				
Capillaries				

Read

SINGLE AND DOUBLE CIRCULATIONS

Recall

1. a) Copy and label figure 11.4.
   b) On your diagrams, show the direction of flow for both oxygenated and deoxygenated blood.

Figure 11.4

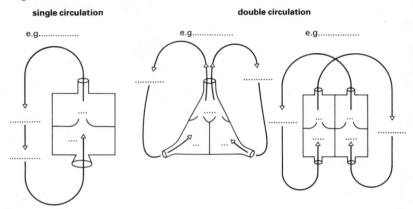

2. List the advantages of a double circulation.

Read

OPEN AND CLOSED CIRCULATIONS

Recall

1. Explain the meaning of (a) open and (b) closed circulatory systems.
2. By means of an annotated diagram, describe the role of the following in blood circulation in a named arthropod:
   (a) haemocoel, (b) the heart, (c) the alary muscles.
3. a) List the functions of insect blood.
   b) State one important functional difference between insect blood and mammalian blood.

Review

1. a) Briefly describe the structure of mammalian blood. [8]
   b) Describe the role of the blood in defending the body against disease. [12]
2. a) Draw a fully labelled diagram to show the circulatory system in a mammal. [8]
   b) Explain briefly how the mammalian circulatory system differs from that of a fish and frog. [4]
   c) Describe the changes that occur in mammalian blood as it passes through (i) the lungs and (ii) the liver. [8]
3. The traces in figure 11.5 are from an electrocardiogram of the heart of a 15-year-old boy. **A** was taken while he was asleep, **B** during normal waking and **C** while running.
   a) Name *two* pieces of apparatus that would have been necessary to obtain these traces. [2]
   b) Indicate differences between trace **A** and trace **C**. [4]

Figure 11.5

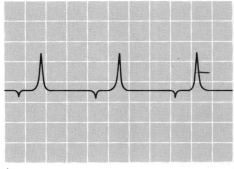

A

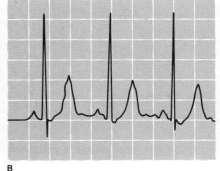

B
C

c) Redraw trace **B** and label the successive phases of a single heartbeat. [4]
d) Describe in some detail how the activity of the heart is controlled so that its output can be varied according to the animal's needs at any given time. [7]
e) Suggest *three* physiological properties that heart (cardiac) muscle might be expected to show which differ from the properties of striated muscle. [3] (AEB)

4 a) By means of annotated diagrams, describe the structure and function of the mammalian heart. [15]
   b) Explain the role of the pacemaker in the mammalian heart. [5]

5 Study figure 11.6 which indicates the variations in pressure in different parts of the human circulatory system. Answer the following questions.

Figure 11.6

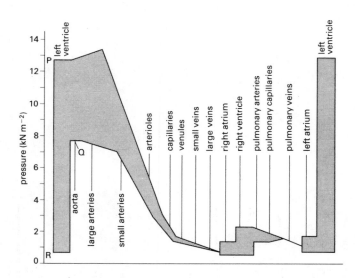

a) Which part of the circulation shows the greatest fluctuations in pressure? [1]
b) Which parts show no fluctuations? [2]
c) Which part of the circulation always maintains a pressure above 7.5 kN m^{-2}? [1]

d) Describe the factors which exert the pressures evident at (i) **P**, (ii) **Q**, (iii) **R**. [6]

e) Suggest the sector of the circulation which offers the greatest resistance to the flow of blood. How is this indicated by the diagram? [3]

f) Blood flow in the main veins is under relatively low pressure. Describe how the return of blood to the heart is maintained. [3]

g) Account for the different ranges of pressures in the pulmonary and systemic circulations. What are the advantages to the mammal in having a double circulatory system? [5]

h) Describe the structural and physiological features of cardiac muscle which distinguish it from skeletal muscle. [4]

(AEB)

12
Uptake and transport in plants

Survey and question
Survey this chapter and make overview notes in pattern form.

Read
UPTAKE OF CARBON DIOXIDE
THE STOMATA
THE MECHANISM OF STOMATAL OPENING AND CLOSURE

Recall
1. Draw an annotated diagram which shows the surface view of a few cells from the underside of a leaf. Include the following in your diagram: (a) epidermal cells, and (b) a stoma showing the structure of the guard cells.

2. a) What is a porometer?
   b) Why does the deflated bulb of a porometer eventually reflate?
   c) What practical precautions need to be taken when using a porometer?

3. The following events are believed to occur in guard cells during *daylight*. Arrange them in the correct sequence.
   a) sugar is produced and accumulates,
   b) stomatal pore opens,
   c) photosynthesis occurs,
   d) sunlight falls on chloroplasts in the guard cell,
   e) guard cell becomes turgid,
   f) OP of guard cell rises,
   g) water passes in to guard cell from adjacent cells.

4. a) Give two reasons why the concentration of sugar in the guard cells decreases at night.
   b) How does this affect the turgidity of the guard cells?

5 a) The opening of stomata is believed to be due to the accumulation of sugar in the cytoplasm and vacuoles of the guard cells. List three reasons why this cannot be the case for all plants.
   b) What alternative hypothesis could explain stomatal opening in these plants?

Read
UPTAKE AND TRANSPORT OF WATER
TRANSPIRATION
MEASURING THE RATE OF TRANSPIRATION
THE EVAPORATING SURFACE
FACTORS AFFECTING THE RATE OF TRANSPIRATION

Recall

1 Define transpiration.

2 Make an annotated diagram to show a qualitative method for observing transpiration from a leaf using cobalt thiocyanate paper. The annotations should include the following information:
   a) the changes you would expect to observe in the thiocyanate paper,
   b) how these changes might differ for the upper and lower surfaces of the leaf, and two possible reasons for these differences.

   (You may find it useful to refer to the information on leaf structure given on page 150 of BAFA)

3 Describe how you could
   a) measure the mass of water lost per unit time per unit leaf area from a potted plant. Suggest a control for this experiment.
   b) In using the results gained from this investigation to estimate the transpiration rate from leaves two assumptions have to be made. What are they?

4 a) Describe how you could measure the volume of water taken up per unit time by a cut shoot.
   b) What assumption is involved if you use the information from this investigation as a measure of transpiration?

5 Transpiration is affected by factors in the external environment of the leaf. Consider a potted plant placed in a stream of warm air (e.g. from a hair-dryer).

a) List three external factors that would most affect its transpiration rate,
b) Explain how each of these factors would affect the transpiration rate,
c) Would you expect an atmometer to be affected in the same way as the plant?

6 a) Define guttation.
   b) Under what circumstances does guttation occur?

7 Explain how transpiration from a leaf and evaporation from an atmometer are affected by (a) a lowering of atmospheric pressure and (b) an increase in light intensity.

8 Apart from environmental conditions, what other factor(s) may influence transpiration?

9 In which direction will water tend to flow through a leaf:
   a) from a region of low osmotic pressure to a region of high osmotic pressure, or
   b) from a region of high osmotic pressure to a region of low osmotic pressure?
   c) Explain why this flow occurs.
   d) Make an annotated diagram to show the flow of water through a leaf from a xylem vessel to the atmosphere. Indicate two possible routes taken by the water through the leaf.

Read

STRUCTURE OF THE ROOT
UPTAKE OF WATER BY THE ROOTS
INTO THE VASCULAR TISSUES OF THE ROOT
FROM ROOT TO LEAF

Recall

1 Make an annotated diagram to show the passage of water from the soil to a xylem vessel in a root. On the diagram, indicate
   a) the different tissue regions across which the water passes.
   b) three possible routes taken by the water across the tissues.

2 a) What evidence is there that transpiration pull alone is not responsible for the upward movement of water between root and stem?
   b) Describe the possible roles of the endodermis in generating root pressure.
   c) What evidence is there to suggest that root pressure involves the expenditure of energy?

Read

STRUCTURE OF THE STEM
XYLEM TISSUE
THE XYLEM AND WATER TRANSPORT
THE ASCENT OF WATER UP THE STEM
THE FUNCTIONS OF TRANSPIRATION
WATER STRESS

Recall

1 By means of annotated diagrams, illustrate the structure of xylem, indicating the relationship between structure and functions of the different types of conducting element.

2 Describe how the following have contributed to our knowledge of the function of xylem:
   a) Eduard Strasburger,
   b) H.H. Dixon,
   c) experiments on cut stems placed in a dye such as eosin.

3 The two sets of apparatus shown in figure 12.1 were set up and left in a dry atmosphere.

Figure 12.1

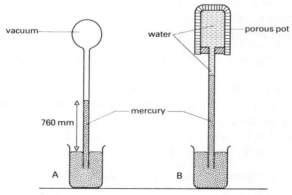

a) Explain why mercury rises up the tube to a height of 760 mm in **A**.
b) Explain why the level of mercury is higher in **B** than in **A**.
c) Why is **B** a useful model for demonstrating the upwards movement of water in the stems of plants?

4 a) What is capillarity?
   b) How could you demonstrate it?
   c) What factor limits the rise of water in a capillary tube?
   d) Why is it unlikely that capillarity plays a significant role in the uptake of water by roots?

5 a) What prevents the columns of water in xylem elements from breaking when subjected to transpiration pull?
   b) List three pieces of evidence which suggest that the ascent of water (sap) in stems may not be simply a physical process, but may also involve living cells.

6 a) List three functions of the transpiration stream.
   b) Explain the meaning of water stress.

Read
UPTAKE OF MINERAL SALTS
SPECIAL METHODS OF OBTAINING ESSENTIAL ELEMENTS

Recall

1 Examine figure 12.13 in BAFA and answer the questions below.
   a) Name the process which probably accounts for these differences in ion concentration.
   b) State two reasons why the differences in ion concentration cannot be explained by diffusion.

2 Examine figure 12.14 in BAFA and answer the questions below.
   a) Explain the differences in ion uptake shown by the plants in the three different sets of conditions.
   b) Explain how these results indicate that ion uptake is not due to diffusion alone.
   c) Describe further evidence which suggests that ion uptake is an active metabolic process.

3 Explain the possible role of the following in ion transport in the root:
   a) plasmodesmata,
   b) cellulose cell walls,
   c) the endodermis.

4 Explain why (a) the mycorrhiza of a conifer and birds nest orchid and (b) a root nodule are examples of symbiosis. In each case, name the organisms involved and the elements (or compounds) which form the basis of the symbiosis.

5 Name one other source of nitrogen which is exploited by plants growing in nitrogen-deficient soils.

Read
TRANSPORT OF ORGANIC SUBSTANCES
ESTABLISHING THE SITE OF TRANSLOCATION
THE MECHANISM OF TRANSLOCATION

Recall

1 Explain what happens to the products of photosynthesis (a) in the spring and (b) later in the year.
2 Define translocation.
3 a) How does the structure of a sieve tube differ from that of a companion cell?
  b) How can these structural differences be related to differences in function? Illustrate your answer with simple annotated diagrams.
4 Figure 12.2 illustrates the results of a ringing experiment carried out by Malphigi in 1679. Suggest a hypothesis to explain these observations.

Figure 12.2

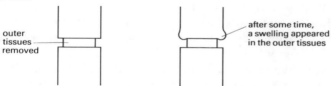

Briefly describe how you could
a) show that the transport of organic materials is confined to the phloem,
b) obtain a sample of pure phloem sap for analysis,
c) calculate the speed of translocation in the phloem.

The following plants lack chlorophyll: dodder, broomrapes, toothwort.
State (a) the name given to the type of nutrition they exhibit,
     (b) the part of the host plant that they attack,
     (c) the host tissue from which they obtain organic nutrients.

7 Munch's model to demonstrate the mass flow hypothesis is shown in figure 12.17 (page 195 in BAFA). Figure 12.3 shows another way of representing the hypothesis, relating it more closely to the structure of the plant. Make a copy of this diagram and, referring at the same time to figure 12.17, answer the following questions.

Figure 12.3

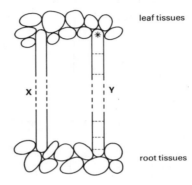

a) State which structures **X** and **Y** correspond to in
   (i) figure 12.16, and
   (ii) a living plant.
b) In a living plant, high sugar concentrations are maintained at *. Explain
   (i) how this is thought to occur, and
   (ii) the effect this has on the osmotic pressure at * and the movement of water (indicate by means of arrows on your diagram).
c) Indicate on your diagram the direction of flow of sap in **Y** and water in **X**. What process maintains the flow of water in **X**?

8 a) State three reasons why the mass flow hypothesis cannot fully explain translocation.
   b) Which of the above reasons has led to the formulation of the elctro-osmotic theory?

9 Briefly outline three alternative hypotheses to the mass flow hypothesis.

Review

1 a) List the substances that are transported in the vascular system of a flowering plant. [4]
   b) Describe one experiment which shows that there is an

upward movement of water through the stem of a plant. [8]
c) Outline two hypotheses which have been proposed to explain the movement of water from root to leaves. [8]

2 a) What is meant by transpiration? How does it (i) resemble, (ii) differ from perspiration and sweating?
   b) In an experiment to compare the rate of transpiration of a runner bean plant with the rate of evaporation of water from a porous pot over a period of 24 hours the following results were obtained:
   The rate of evaporation was measured in millilitres per hour. (ml h^{-1}).
   The rate of transpiration was measured in millilitres per square metre of leaf surface per hour. (ml m^{-2} h^{-1}).

Table 12.1

Time of day	Rate of evaporation (ml h^{-1})	Rate of transpiration (ml m^{-2} h^{-1})
6–8 a.m.	3.9	93
8–10 a.m.	6.7	163
10–12 noon	8.2	222
12–2 p.m.	9.5	253
2–4 p.m.	9.5	198
4–6 p.m.	9.1	181
6–8 p.m.	6.7	126
8–10 p.m.	3.9	8
10–12 midnight	3.4	19
12–2 a.m.	1.6	19
2–4 a.m.	0.8	14
4–6 a.m.	0.9	24

By using appropriate scales on the same squared paper, plot graphs of the rate of evaporation from the porous pot and the rate of transpiration of the bean plant against time of day.
Comment on the significance of the shapes of the graphs. [28]
(Oxford)

3 a) Give an illustrated account of the movement of water from the soil into the xylem cell of the root. [10]
   b) Describe the role of the endodermis in the movement of water across the root. [6]
   c) Outline the experimental evidence for this role. [4]

4 A liquid fertilizer containing radioactive phosphorus (^{32}P) was sprayed onto the soil in a plot of natural grassland. The radioactive phosphorus was taken up by the plants.

The graph shows the relative levels of radioactivity in three types of organism living in the plot. Readings were taken at intervals over a period of 35 days.

Figure 12.4

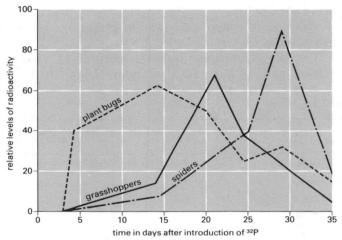

a) Comment on the biological significance of the initial build up of radioactivity in each of the three types of organism. [6]
b) Comment on the maximum levels of radioactivity reached in each type of organism. [3]
c) Explain why the level of ^{32}P falls rapidly towards the end of the experiment. [1]
d) Describe the procedures which could be used to obtain the data after the initial application of ^{32}P. [4]
e) What implications can be drawn from the data concerning the dangers of radioactive materials which persist in the environment? [1]
f) Construct a food web to illustrate the feeding relationships which probably exist in such an area of grassland. Include the three organisms featured on the graph. [5]   (AEB)

5 Young seedlings were allowed to absorb radioactive isotopes of calcium and phosphorus (incorporated in a soluble ion) for one hour. They were then removed to a non-radioactive solution and were sampled 5 minutes, 6 hours and 48 hours after the radioactive treatment.

Figure 12.5

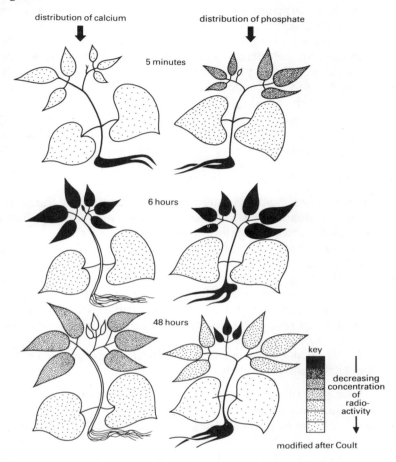

The shading on the diagrams (figure 12.5) is directly proportional to the amount of radioactivity found to be present in any one part of the plant.
a) Compare the distribution of calcium and phosphate after 5 minutes. [4]
b) Compare the distribution of calcium and phosphate after 6 hours. [4]
c) Compare the distribution of calcium and phosphate after 48 hours. [4]
d) Make a general comparison between the mobility of the calcium and phosphate in the seedlings. [4]

92 Study Guide

   e) Correlate one universal function of phosphate in living tissue with the distribution in this experiment. Clearly specify this function. [4]
   f) The results with these seedlings are also characteristic of the behaviour of calcium and phosphate in a deciduous tree. Draw a calcium cycle and a phosphate cycle, indicating the similar and dissimilar features of the recycling of these two ions which would take place in a deciduous wood. [6]
   g) Describe an experiment using these isotopes and woody privet shoots to determine the extent to which these elements are translocated in the phloem or the xylem. [4] (AEB)

6 a) Consider the statements given below:

There is an endogenous (internal) rhythm controlling the movement of stomata which results in their opening during the hours of daylight and closing during the hours of darkness.

Stomata show an opening response to a reduction in the carbon dioxide concentration in the sub stomatal chamber below the normal 0.3% present in atmospheric air, and a closing response with an increase in carbon dioxide concentration.

The rates of respiration and of photosynthesis increase with a rise in temperature, but the rate of respiration increases more rapidly than the rate of photosynthesis.

Lack of water in the mesophyll cells of a leaf leads to a reduction in the rate of photosynthesis.

Using the information given above and your own knowledge of photosynthesis explain why:
   (i) the rate of photosynthesis rises dramatically at daybreak; [4]
   (ii) high temperatures bring about the closure of stomata during the hours of daybreak; [2]
   (iii) an increase in wind speed can result in the closure of stomata during daylight; [2]
   (iv) an increase in the amount of carbon dioxide above 1% leads to a reduction in the rate of photosynthesis. [1]
   b) Give a brief account of the probable mechanism causing the movement of guard cells and which leads to the opening of stomata. [6]
   c) Describe a method you would use to accurately determine the distribution of stomata on a leaf. [5] (AEB)

13
The principles of homeostasis

This is the first chapter in part III: 'Adjustment and control'. The overview notes below show how the topics covered by the chapters in this part of the book are related.

Overview notes for part III: Adjustment and control
(Chapter numbers given in brackets)

Figure 13.1

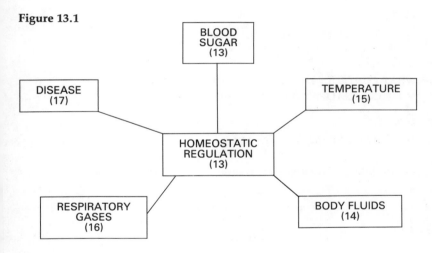

Survey and question
Survey this chapter and make overview notes in pattern form.

Read
THE MEANING OF THE INTERNAL ENVIRONMENT
THE FORMATION OF INTERCELLULAR FLUID
WHAT FACTORS MUST BE KEPT CONSTANT?

Recall

1. **a)** What is meant by the internal environment?
   b) By means of an annotated diagram, describe how intercellular fluid is formed in mammals. On your diagram show the routes by which intercellular fluid is returned to the blood circulation.
   c) List the functions of intercellular fluid.

2. **a)** Describe some experimental evidence which indicates the importance of homeostasis.
   b) List those aspects of the internal environment that are regulated homeostatically.

3. What is the meaning of the following terms: (a) frog Ringer's (solution) and (b) invertebrate saline.

Read

THE HOMEOSTATIC CONTROL OF GLUCOSE
THE ROLE OF THE PANCREAS
DIABETES
WHAT CONTROLS THE SECRETION OF INSULIN?
SOME GENERALIZATIONS ABOUT HOMEOSTASIS

Recall

1. Copy and complete figure 13.2.

Figure 13.2 *The control of blood sugar levels by pancreatic hormones*

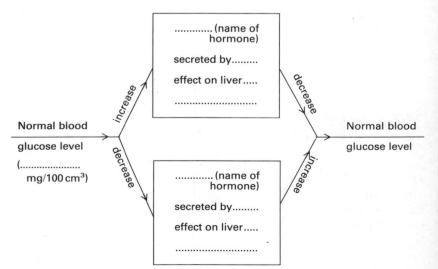

2 a) State two causes of hyperglycaemia.
   b) What are the symptoms of hyperglycaemia?
   c) How may diabetes melitus be controlled?
3 a) Add to your diagram (from question 1) further notes about the control of insulin secretion.
   b) This diagram (from question 1) summarises the principles of a homeostatic control mechanism. Add to your diagram the following elements of a homeostatic control process: receptors, effectors, negative feedback, corrective mechanisms.

Read
THE MAMMALIAN LIVER
FUNCTIONS OF THE LIVER

Recall
1 As shown in table 13.1, tabulate the following functions of the liver with brief explanatory notes: carbohydrate metabolism, protein metabolism, lipid metabolism, detoxification, production of heat and bile, elimination of waste materials, storage.

Table 13.1

Function	Notes
Carbohydrate metabolism etc.	Regulation of sugar Liver cells convert glucose to glycogen or break it down to CO_2 and H_2O etc.

Read
THE STRUCTURE OF THE LIVER

Recall
1 a) Name the blood vessels which carry blood to and from the liver.
   b) List the differences in the constituents of the blood in each of these vessels.
2 Why is the liver so well supplied with blood?
3 a) Name the cells which carry out the functions of the liver.
   b) Describe the structure of one of these cells.

Figure 13.3

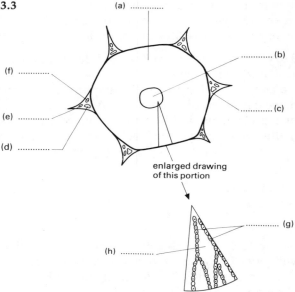

4 Figure 13.3 is a drawing of a photomicrograph of a section through a mammalian liver. Make a copy of the drawing, name the labelled parts (a–h) and write short notes on each.

Review

1 a) What is homeostasis? [4]
   b) Describe the role of each of the following in homeostasis.
      (i) the liver.
      (ii) the pancreas. [8, 8]

2 Give an account of the methods by which a constant internal environment is maintained in a mammal. [20]

3 With reference to a named biochemical or physiological process, explain the meaning of:
   a) a homeostatic control mechanism. [10]
   b) positive feedback. [5]
   c) negative feedback. [5]

14

Excretion and osmoregulation

Survey and question
Survey this chapter and make overview notes in pattern form.

Read
THE MAMMALIAN KIDNEY
THE STRUCTURE OF THE NEPHRON

Recall
1 Define (a) excretion, (b) nitrogenous excretion and (c) osmoregulation.
2 Name the organs responsible for excretion and osmoregulation in a vertebrate.
3 a) Copy figure 14.1 and label it.

Figure 14.1 *Diagram of a mammalian nephron*

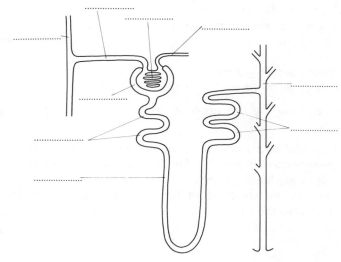

98 Study Guide

    b) Note that the diameter of the afferent blood vessel leading to the glomerulus is greater than that of the efferent vessel leading away from it. How will this affect the speed of blood flow through the glomerulus?

Read
FILTRATION IN BOWMAN'S CAPSULE

Recall
1. **a)** Name the process which occurs in Bowman's capsule.
   b) Describe the experimental evidence which led to an understanding of the functions of the Bowman's capsule.
   c) List the components of glomerular filtrate.
2. **a)** State the requirements necessary for ultrafiltration.
   b) Describe the structural evidence that indicates that the Bowman's capsule and the glomerulus fulfil one of these requirements.
   c) Explain how the second requirement is fulfilled.

Read
REABSORPTION IN THE TUBULES
FINE STRUCTURE OF THE TUBULE CELLS

Recall
1. Refer back to your drawing of a nephron. Indicate on it where each of the following occurs:
   a) absorption and active transport of glucose,
   b) absorption and active transport of salt (sodium and chloride ions),
   c) absorption of water,
   d) collection of urine.
2. Explain how the functions of the kidney tubules were investigated.
3. **a)** Describe two lines of evidence which suggest that active transport is responsible for the reabsorption of glucose and salt in the kidney tubule.
   b) How is the structure of the kidney tubule epithelial cells related to their function?

Read
THE ROLE OF THE LOOP OF HENLE

Recall
1 Copy figure 14.7 in BAFA and label it. Add the annotations below to explain the functions of the loop of Henle:
   a) Thick walled tubule, impermeable to outward movement of water, but not salt.
   b) Outward movement of salt due to active transport of sodium chloride into the surrounding medullary tissue. (Indicate on your diagram that this outward movement occurs in all directions.)
   c) A high concentration of salt builds up in the medullary tissue. Bloodflow here is sluggish, and therefore the high salt concentration is maintained. This, together with urea retention by these tissues, helps build up a high osmotic pressure in the medullary tissue.
   d) Salt concentration in the medullary tissue is highest at the apex of the loop.
   e) Sodium ions are absorbed from the medullary tissue into the descending limb of the tubule.
   f) As fluid flows down the descending limb, it increases in salt concentration; as it flows up the ascending limb it becomes more dilute. Why?
   g) Fluid passes down the collecting duct through medullary tissue of increasing salt concentration. Water passes out of the collecting duct by osmosis into the surrounding tissues.

Read
THE KIDNEY AS A REGULATOR

Recall
1 The list below summarises the regulatory events that occur when excessive amounts of water are lost from the body or a large quantity of salt is eaten. Arrange these events in the correct sequence.
   a) Water passes out of the kidney tubules into the surrounding blood vessels and enters the general circulation.
   b) Antidiuretic hormone (ADH) is released into the bloodstream by the posterior lobe of the pituitary gland.
   c) Osmoreceptor cells in the brain (hypothalamus) are stimulated by a rise in the osmotic pressure of the blood.

d) ADH acts on cells lining the distal convoluted tubule and the collecting duct, making them more permeable to the outward passage of water.
e) The body loses water (or gains salt) so the body fluids become more concentrated. The osmotic pressure of the blood rises.
f) ADH is secreted by specialised nerve cells in the hypothalamus and flows from here to the pituitary via nerve axons.

2 Describe and explain the symptoms of diabetes insipidus.

Read
EXCRETION AND REGULATION IN OTHER ANIMALS
MARINE INVERTEBRATES
FROM SEA TO FRESHWATER
CONTRACTILE VACUOLE
ANTENNAL GLANDS

Recall
1 Explain the general principle illustrated by each of the following examples:
   a) A fresh water fish dies if placed in sea water.
   b) *Maia* dies if placed in fresh water.
2 How are the following able to tolerate fluctuations in the osmotic pressure of their environment (OPe)?
   a) *Eriocheir*   b) *Arenicola*   c) *Hydrobia*.
3 a) In fresh water animals, OPi is greater than OPe. How do you account for this?
   b) Why is it necessary for these animals to possess an osmoregulatory mechanism?
4 Describe the evidence which suggests that the contractile vacuole in *Amoeba* is osmoregulatory and utilises energy.
5 Account for each of the following.
   a) Fluid from the end sac of *Carcinus* is isotonic with its blood, and an isotonic urine is produced (i.e. salts are lost in the urine). The blood of *Carcinus* has an OP greater than that of sea water.
   b) Fluid from the end sac of a fresh water crayfish is isotonic with its blood, but a hypotonic urine is produced. The blood of a fresh water crayfish has an OP greater than that of fresh water.

Excretion and osmoregulation 101

Read
FRESH WATER FISHES
FROM FRESH WATER BACK TO THE SEA
MARINE ELASMOBRANCHS
MIGRATORY FISHES

Recall
1 Make a copy of figure 14.12 in BAFA and annotate the drawings fully to show how fresh water and sea water teleosts maintain osmotic stability.

2 Make a similar diagram to figure 14.12 in BAFA, adapted to show how marine elasmobranchs maintain osmotic stability.

3 Suggest possible mechanisms by which migrating fish (e.g. salmon and eels) are able to osmoregulate successfully in conditions which vary from one extreme of osmotic environment to another.

Read
FROM FRESH WATER TO LAND

Recall
1 Table 14.1 summarises some of the adaptations by which terrestrial animals conserve water. Copy and complete it.

Table 14.1

Adaptation	Named example(s)	Explanation of adaptation
Reduction of water loss from outer surface	Insects etc.	Cuticle has an outer layer of wax etc.
Reduction of water loss by excretory organs		
Behavioural pattern which reduces water loss		
Physiological and metabolic adaptations		

Read
OSMOREGULATION IN PLANTS

Recall

1. **a)** Copy and complete table 14.2, which summarises the ways in which the roots, stems and leaves of xerophytic plants may be adapted to their mode of life.

    Table 14.2 *Adaptations of xerophytes*

Roots	Stems	Leaves

    b) Complete the following:
    Mesophytes live in ....
    Xerophytes live in ....
    Hydrophytes live in ....
    Halophytes live in ....

2. **a)** Name the process by which plants lose water.
    b) Under what conditions is water loss greatest?
    c) Describe how this water loss occurs.

Read
CONTROL OF IONS

Recall

1. **a)** Make two copies of figure 14.2. Substitute 'sodium and potassium' for 'ION' in the first diagram and 'calcium and phosphate' for 'ION' in the second. Complete the diagrams by filling in the details of the corrective mechanisms in each case.

Figure 14.2 *Summary of the control of ion concentration in the blood of a mammal*

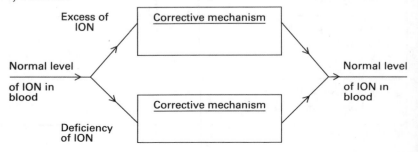

b) What controls the production of aldosterone by the adrenal cortex?

2 Make brief notes on ionic regulation in plants.

Review

1 a) What is osmoregulation? [2]
   b) Explain how osmoregulation is carried out in:
      (i) a named fresh water protozoan,
      (ii) a named crustacean,
      (iii) a named marine fish. [4, 6, 8]

2 a) Distinguish clearly between excretion and secretion. [2]
   b) List the excretory products of a mammal, an insect and a green plant. [6]
   c) Explain how nitrogenous excretion is carried out in an insect, a mammal and a green plant. [12]

3 a) Make fully labelled diagrams to show:
      (i) the general structure of the mammalian kidney.
      (ii) the structure of a nephron and its blood supply. [10]
   b) Describe the main functions of each region of the nephron. [6]
   c) Explain how the kidney functions as a homeostatic device. [4]

4 Table 14.3 gives information about substances which pass through normal human kidneys.

Table 14.3

Substance	Approximate daily quantities	
	filtered through the glomeruli	excreted
Water	170 l	1.5 l
Sodium	547 g	2.3 g
Glucose	187 g	None
Amino acids	8.5 g	0.15 g
Urea	51 g	30 g

a) The kidney plays an important role in the elimination of waste compounds containing nitrogen from the body.
What do the data in the table indicate about the form in which this nitrogen is eliminated? [2]

b) It is evident from the table that glucose retention is normally 100%. Suggest *three* different physiological conditions under

which glucose may be found in the urine. Explain why these conditions cause loss of glucose from the body. [6]
c) Describe (i) a biochemical test which could be used to confirm that no glucose is excreted and (ii) a biochemical test to show that amino acids are present in only very small quantities. [4]
d) Calculate to one decimal place the percentages of water and sodium which are retained in the body. [2]
e) From your knowledge of kidney function, compare the processes by which water and sodium are reabsorbed. [10]
f) Using the information in the table, the fact that 80% of filtered water is reabsorbed in the proximal convoluted tubule, and your own knowledge, construct a three-bar histogram to show the number of litres of water reabsorbed in
   (i) the proximal convoluted tubule,
   (ii) the loop of Henle,
   (iii) the distal convoluted tubule and the collecting ducts. [6]
(AEB)

5 Table 14.4 gives data about the composition of blood plasma and urine for a man.

Table 14.4

	Plasma	Urine
Water	90%	96%
Proteins	7%	0%
Hormones	trace	trace
Ions	0.9%	Variable, usually more sodium, chloride, sulphate than in plasma
Waste substances	trace	2%
Glucose	0.1%	0%
pH	7	varies, usually 6

a) What do you deduce about the function of the kidney from information provided by the table? Illustrate your answer by reference to specific differences in the composition of plasma and urine. [6]
b) Assuming that renal function is normal, what are the *two* main conditions which are responsible for varying the mineral ion concentration of the urine? [2]
c) Given that the salt content of sea water is 3% and the maximum content of salts in human urine is 2.2%, attempt to

explain why a shipwrecked sailor is unable to survive for long by drinking sea water. [2]
d) How do the kidney and the brain function together to bring about a stable osmotic environment in the body? [4]
e) How would you expect the composition of human urine to differ from that indicated in the table above:
   (i) after strenuous exercise,
   (ii) during a high protein diet? [6]   (AEB)

6 The graphs in figure 14.3 show the relationship between the osmotic concentration of the blood and the external medium in three invertebrate animals. The osmotic pressure is expressed as depression of the freezing point (Δ°C).

Figure 14.3

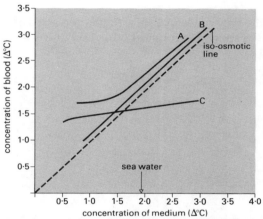

a) Examine the graphs and then suggest a likely habitat for each organism, giving reasons for your choice. [6]
b) (i) Explain why fresh water teleosts (bony fish) excrete large quantities of dilute urine, whereas marine teleosts excrete small amounts of more concentrated urine.
   (ii) Marine elasmobranchs (cartilaginous fishes) retain an organic compound in their blood which raises its osmotic concentration to that of sea water. What volume and concentration of urine may they be expected to excrete when compared with the two types of teleosts described above? [8]
c) Carefully explain how you would carry out an experiment to investigate the ability of a soft-bodied animal such as a marine worm to regulate the concentration of its body fluids. [6]
       (AEB)

15
Temperature regulation

Survey and question
Survey this chapter and make overview notes in pattern form.

Read
TEMPERATURE REGULATION

Recall
1 a) Name the type of temperature regulation shown by **A** and **B**.
   b) What is responsible for supplying body heat to **A** and **B**?

Figure 15.1

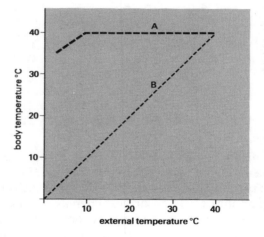

2 a) What are the advantages of maintaining a constant body temperature?
   b) State two ways by which a constant body temperature may be maintained.

Read
HOW HEAT IS LOST AND GAINED
RESPONSE TO COLD BY AN ENDOTHERMIC ANIMAL
RESPONSE TO HEAT BY AN ENDOTHERMIC ANIMAL

Recall

1 Copy table 15.1 and complete it by describing how radiation, evaporation, conduction and convection are either prevented by or made possible by physiological mechanisms in endothermic animals.

Table 15.1

Physical process of heat transfer	Prevented by (a response to cold)	Made possible by (a response to heat)
Radiation	Thick layer of sub-cutaneous fat ...	
Evaporation	etc.	
Conduction		
Convection		

Read

THE ROLE OF THE BRAIN IN TEMPERATURE REGULATION

Recall

1 Copy figure 15.2 and complete it by writing in the corrective mechanisms.

Figure 15.2 *Homeostatic control of body temperature in a mammal*

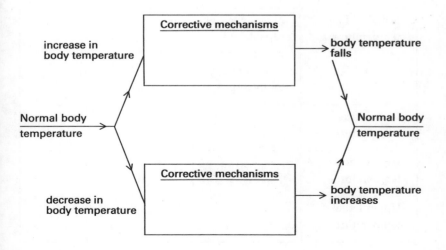

2 a) Name the temperature control centre in mammals.
   b) Describe the experimental evidence which indicates that this centre is responsible for changes in the temperature of the blood.
   c) What role do the thermoreceptors in the skin play in temperature control?

Read

THE EFFECT OF RAISING AND LOWERING ENVIRONMENTAL TEMPERATURE
TEMPERATURE REGULATION AND THE ENVIRONMENT
BEHAVIOURAL CONTROL OF BODY TEMPERATURE
TEMPERATURE AND HIBERNATION
TEMPERATURE TOLERANCE

Recall

1 a) Explain the meaning of the body's efficiency range.
   b) Copy and complete table 15.2.

Table 15.2 *A summary of the effects of environmental temperature on body temperature control in mammals*

Environmental temperature	Effect on physical mechanisms of temperature control	Effect on biochemical mechanisms of temperature control	Effect on body
Lower lethal temperature	No longer capable of maintaining constant body temperature	etc.	etc.
Low critical temperature			
High critical temperature			
Upper lethal temperature			

2 Explain the significance of the following.
   a) The arctic fox has a low critical temperature of −40 °C whereas the kangaroo rat has a low critical temperature of 31 °C.
   b) The smallest endothermic vertebrates tend to be restricted to warm regions, whilst the larger endothermic vertebrates tend to be found in colder regions.
   c) The ears of elephants have a rich blood supply.
   d) In the flippers of dolphins, the arteries and veins are very close together.
   e) Desert lizards are more active in the early morning and evening.
   f) Swallows migrate southwards in winter.

3 Draw graphs to show the relationship between body temperature and environmental temperature in
   a) a named hibernating endotherm throughout one year,
   b) a named non-hibernating endotherm throughout one year,
   c) a camel throughout a 24 hour cycle,
   d) a humming-bird throughout a 24 hour cycle.

Read
TEMPERATURE CONTROL IN PLANTS

Recall
1 a) Name the process by which plants keep cool.
   b) Describe how this process brings about cooling.
   c) Explain how wilting can be important in temperature regulation in plants.

2 Using a named example, describe how plants can keep cool in very hot climates.

Review
1 a) Make a fully labelled diagram of a vertical section through the human skin to show the structures visible under a light microscope. [8]
   b) Discuss the role of the skin in the control of body temperature in a mammal. [12]

2 a) List four ways by which mammals can reduce their rate of heat loss to the environment. [4]
   b) Explain the physiological mechanisms underlying each of them. [10]

c) Briefly describe the role of the mammalian brain in temperature regulation. [6]

3 The heart of an aquatic annelid worm was observed through its transparent dorsal surface to investigate the effect on it of temperature. The time taken for 10 heart beats was recorded, first at room temperature, 20 °C, and then at 2 °C intervals as the water was gradually cooled. The worm was then returned to room temperature and allowed to stabilise. Further recordings were taken again as the temperature was gradually raised to 30 °C. Finally the worm was replaced in water at room temperature and another reading taken.

Reading number	1	2	3	4	5	6	7	8	9	10	11
Temperature °C	20°	18°	16°	14°	20°	22°	24°	26°	28°	30°	20°
Time (in seconds) taken for 10 beats	39	52	58	67	40	39	32	20	47	70	86

a) Plot the points of rate of heart beat against temperature on graph paper. Join these points in a suitable way to indicate what is happening to the worm.

b) What would you expect to happen to the heart rate, if the animal were subsequently:
   (i) maintained at 20 °C,
   (ii) lowered once more to 14 °C,
   (iii) raised finally to 30 °C?
   Give your reasons. (Oxford)

4 Discuss the adaptations of animals living in deserts. [20]

5 Figure 15.3 summarises the ways in which an animal can gain heat from or lose heat to its environment.
   a) With the aid of the information provided in the diagram, explain how an ectothermic animal (i.e. one with a large surface area which has a high rate of exchange with the environment) such as a lizard can
      (i) raise its body temperature to that of the environment,
      (ii) lower its body temperature below that of the environment. [8]
   b) Carefully explain how an increase in air currents can influence the amount of heat a mammal loses to the environment. [2]
   c) Explain the connection between problems of heat regulation

Figure 15.3

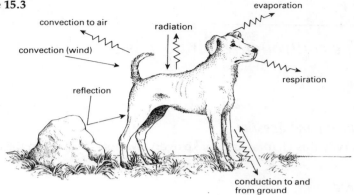

and the fact that there are no adult birds or mammals less than two grams in mass. [3]

d) Heat loss from the skin of a person sitting in an environment at a temperature above that of his body drops soon after he drinks a large mug of iced water.
   What does this information tell you about the way in which the thermoregulatory centre in the hypothalamus region of the brain functions? Explain your answer. [4]

e) When the body temperature of the marine crustacean *Ligia* approaches too high a level it moves out of its normal damp environment beneath stones into direct sunlight and a higher environmental temperature. Explain how this apparently strange behaviour benefits the animal. [3]   (AEB)

16

The control of respiratory gases

Survey and question
Survey this chapter and make overview notes in pattern form.

Read
THE ROLE OF BREATHING AND THE CIRCULATION
THE EFFECTS OF FLUCTUATIONS IN OXYGEN AND CARBON DIOXIDE
CARBON DIOXIDE AS THE STIMULUS IN THE CONTROL PROCESS

Recall
1 a) Make two copies of figure 16.1. Fill in the spaces in the diagrams so that one summarises the control of carbon dioxide, and the second summarises the control of oxygen.

Figure 16.1 *Diagram summarising the homeostatic control of carbon dioxide/oxygen**

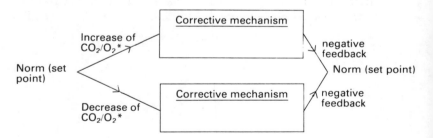

*Delete whichever does not apply.

   b) Describe how oxygen poisoning can occur and its effects on the body.

2 Define (a) anoxia and (b) apnoea.

3 How would you demonstrate experimentally that an increase in CO_2 level in the blood stimulates ventilation?

4 Why is a high level of CO_2 in the blood dangerous?

Read
THE ROLE OF THE BRAIN
CONTROL OF BLOOD PRESSURE
CEREBRAL CONTROL OF RESPIRATION AND CIRCULATION

Recall

1 a) Copy figure 16.2. Annotate it to show the role of the brain in controlling the amount of carbon dioxide in the blood. Label the structures numbered 1 to 6 and indicate how they are affected by an increase in the amount of carbon dioxide in the blood.

Figure 16.2 *Summary of the role of the brain in the control of carbon dioxide levels in the blood*

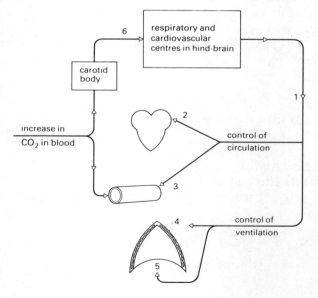

b) Describe the role of the carotid sinus in the control of blood pressure.

2 Add annotations to your diagram to show how
   a) Ventilation can be controlled voluntarily,
   b) circulation is affected by hormones.

Read
ADJUSTMENT TO HIGH ALTITUDE
ADJUSTMENTS DURING EXERCISE

Recall
1 How does the body respond to
   a) gradual decreases and
   b) sudden decreases in the availability of oxygen?
2 Below is a list of the processes that occur in the human body during a sprint. Arrange them in the correct order and write brief notes on each stage.
   a) Rise in body temperature
   b) Stretch receptors in the carotid sinus are stimulated
   c) Local dilation of arteries
   d) Adrenaline is secreted
   e) Muscles start respiring anaerobically
   f) Metabolic rate increases
   g) Panting occurs
   h) Stretch receptors in the aortic and carotid bodies are stimulated.

Read
RESPONSE TO TOTAL OXYGEN DEPRIVATION

Recall
1 a) Define bradycardia.
   b) Describe its effect on the body.
   c) Under what circumstances does bradycardia occur in animals?

Review

1. a) Describe the conditions under which changes are brought about in the rate of heartbeat and ventilation in a mammal. [6]
   b) Explain how the oxygen concentration of the blood is maintained under these conditions. [7]
   c) Under normal circumstances, how are the rate of heartbeat and ventilation kept constant? [7]

2. a) Write an account of the carriage of respiratory gases in the blood of a mammal. [12]
   b) Describe how the level of gases in the blood is controlled. [8]

3. a) Graph 1, figure 16.3, shows the percentage of haemoglobin associated with oxygen to form oxyhaemoglobin over a range of partial pressures of oxygen. Graph 2 shows the relationship between altitude and partial pressures of oxygen.

Figure 16.3

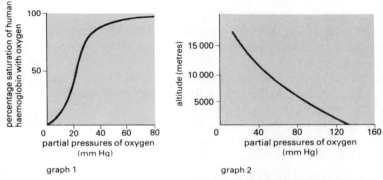

graph 1   graph 2

Note that chemical details of glycolysis and Krebs TCA cycle are **not** required in any part of this answer.

   (i) Using the information given on both graphs explain why most people who are not acclimatised to living at high altitudes will lose consciousness at altitudes between 6000 and 8000 metres. [5]
   (ii) Permanent human habitations occur up to approximately 7000 metres and people who are acclimatised to high altitudes can survive for a few hours when breathing air at approximately 9000 metres.
   Suggest *three* adjustments which probably occur in the physiology of such acclimatised people. [3]
   (iii) Explain the physiological reasons for each of the adjustments you have suggested. [3]

b) The following data refer to Olympic Games held at the sites stated.

Results of the 10 000 m race

Tokyo 1964	Mexico 1968
(200 m above sea level)	(2242 m above sea level)
1. M. Mills, U.S.A.	1. N. Tamu, Kenya**
2. M. Gammoudi, Tunisia**	2. M. Wolds, Ethiopia**
3. R. Clarke, Australia	3. M. Gammoudi, Tunisia**
4. M. Wolds, Ethiopia**	4. J. Martinez, Mexico**
5. L. Ivanov, Russia	5. N. Sviridov, Russia*
6. K. Tsuduroya, Japan	6. R. Clarke, Australia*
7. M. Halberg, New Zealand	7. R. Hill, U.K.*
8. A. Cook, Australia	8. W. Masresha, Ethiopia**

** Indicates athletes who had lived most of their life at high altitudes.

 * Indicates athletes who trained at high altitudes for an extended period prior to the games.

Carefully explain why unacclimatised athletes were relatively unsuccessful during the 10 000 m race at the Mexico Olympic games. [4]
c) Some unsuccessful athletes collapsed and were given oxygen. Clearly explain the role of this oxygen with specific reference to the athletes' livers. [4]
d) By referring only to general principles explain the role of oxygen in energy release in mitochondria. [6]
e) Describe a quantitative experiment which you have carried out to compare the composition of inspired and expired air in any *named* living organism. [5]   (AEB)

17
Defence against disease

Survey and question
Survey this chapter and make overview notes in pattern form.

Read
PREVENTING ENTRY
PHAGOCYTOSIS

Recall
1 Copy and complete figure 17.1.

Figure 17.1 *Micro-organisms and disease*

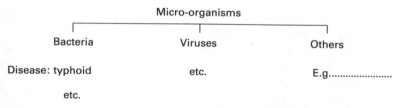

2 Describe some of the effects that invading micro-organisms have on the body.

3 Name two main types of mechanism which defend the mammalian body against disease.

4 a) Name two major pathways through which micro-organisms enter the body.
  b) Describe how the entry of micro-organisms via each of these routes is to some extent prevented.

5 a) What feature of the skin enables it to serve as an effective barrier to the entry of most micro-organisms?
  b) How can this barrier be made ineffective?
  c) Explain briefly how the entry of micro-organisms into the blood system is prevented.

6 a) What is meant by pasteurisation?
   b) Briefly outline how artificial measures can be used to prevent infection.

7 a) Figure 17.2 summarises the events that occur during an inflammation. Make a copy of it and complete the labelling by selecting from the list below.

Figure 17.2 *Inflammation. The skin has been wounded and micro-organisms have invaded the wound. Vertical section through inflammation, not drawn to scale.*

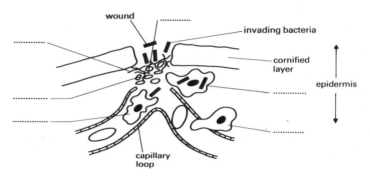

   (i) A phagocyte migrating to the site of the wound
   (ii) Tissue containing phagocytes
   (iii) Network of fibrin forming – traps r.b.c.s. and bacteria, dries forming scab
   (iv) Phagocyte undergoing diapedesis
   (v) Pus formed from dead bacteria and macrophages collecting beneath scab
   (vi) Clot formed by platelets
   (vii) Phagocyte engulfing bacteria

   b) Make brief notes on the role of macrophages in the defence of the body against infection.

Read

ANTIBODY FORMATION
LYMPH NODES
IMMUNITY
THE ROLE OF THE THYMUS GLAND
ALLERGY AND STRESS

Recall

1. **a)** Copy the following passage and complete it by filling in the blank spaces.

   Micro-organisms, like other organisms, contain macromolecules such as carbohydrates, proteins and so on. These are called .... When micro-organisms invade the mammalian body, it is able to 'recognize' these .... as being 'foreign', and certain body cells called .... respond by producing .... This response is called the .... and provides the body with defence or ....

   b) Describe six different methods by which invading micro-organisms may be destroyed by the mammalian body.

2. Copy figure 17.6 in BAFA and annotate it to describe the roles of zones **A**, **B** and **C**.

3. Explain briefly the essential difference between active immunity and passive immunity.

4. Copy and complete table 17.1.

Table 17.1 *A comparison of methods of artificial immunity*

	Vaccine (Active immunity)	Serum (Passive immunity)
Method of obtaining		
Named examples of diseases for which immunity is provided		
Effect on recipient		
Duration of immunity		

5. Describe the effect of removing the thymus gland from (a) new-born mice and (b) older mice.

6. **a)** Describe the evidence which indicates that the thymus exerts its effects by producing a hormone.

   b) List the functions of the thymus.

7 a) List the symptoms of an allergy such as hay fever.
   b) Name the chemical believed to be responsible for these symptoms.
   c) Explain how the release of this chemical into the body is believed to occur.
   d) Describe how allergies can be treated.

8 Make brief notes to describe how the body combats stress.

Read
BLOOD GROUPS

Recall

1 List the ABO blood groups and explain why they are so called.

2 a) Copy table 17.2 and complete it by writing in either 'agglutination' or 'no agglutination', 'yes' or 'no' for advisability of transfusion and antibody/antigen details according to the key.

   Key
   A, B, AB = antigens on red blood cells
   a, b = antibodies in plasma
   O, o = absence of antigens or antibodies

Table 17.2

Donor			Recipient			Agglutination/ no agglutionation	Advisability of transfusion
Group	Antibody	Antigen	Group	Antibody	Antigen		
A			B				
O			A				
AB			A				
A			AB				

   b) With reference to your completed table, explain why people with blood group O can be described as universal donors and those with blood group AB as universal recipients.

3 Copy table 17.3 and complete it by writing in 'Rh' for Rhesus antigens and —— for absence of antigen or antibody.

Defence against disease 121

Table 17.3 *The occurrence of Rhesus antigens*

	Antigen	Antibody
Rhesus positive blood		
Rhesus negative blood		

4 a) Describe the changes which occur in a Rhesus negative individual after receiving Rhesus positive blood.
   b) Explain how this can occur naturally and why it can be so dangerous.
5 How may erythroblastosis foetalis be prevented?

Read
HOW ARE ANTIBODIES PRODUCED?
REJECTION OF GRAFTS
TRANSPLANT SURGERY
INTERFERON

Recall
1 Copy figure 17.9 in BAFA and annotate it to describe the essentials of the instructive hypothesis and the clonal selection hypothesis.
2 a) Which of the two hypotheses better explains why normally we catch certain infective diseases (e.g. mumps) only once.
   b) Explain why the alternative hypothesis does not account for the enhanced immune response mentioned in question 2.a) above.
3 a) What is meant by rejection?
   b) How does rejection differ from the immune response to invading micro-organisms?
4 Describe the difference between (a) immunological tolerance and (b) immuno-suppression.
5 Describe two methods each of inducing (a) immunological tolerance and (b) immuno-suppression and explain the mechanism of the methods.
6 a) What is interferon?
   b) Describe how interferon could be used to protect man against infectious diseases.
   c) Outline the difficulties of producing interferon commercially.

Read

CHEMOTHERAPY AND ANTIBIOTICS

Recall

1 Copy and complete table 17.4.

Table 17.4 *Chemotherapeutic agents*

	Naturally occurring antibiotics e.g.	Synthetically produced e.g. ......
Active against		
Mode of action		

2 a) How was penicillin discovered?
  b) How do modern methods of taking penicillin differ from earlier ones?
  c) Describe one serious drawback of chemotherapeutic agents.

Review

1 Describe how each of the following could be used to reduce the incidence of human disease.
  a) immunisation, [5]
  b) chemotherapy, [5]
  c) interferon, [5]
  d) transplantation surgery. [5]

2 a) What are antibodies? [4]
  b) The graphs in figure 17.3 show the results of an experiment on antibody production in mice.

Figure 17.3 *Antibody production during the course of a primary (A) and secondary (B) response*

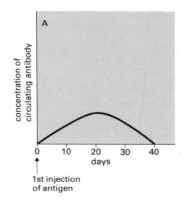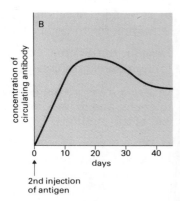

Explain the reasons for the differences in response shown between **A** and **B**. [6]
c) How are antibodies believed to be produced by the body? [10]

3 a) Distinguish between (i) active and passive immunity, and (ii) antibody and antigen. [4, 4]
   b) Account for the fall in the death rate throughout the world of the following diseases: malaria, bubonic plague and smallpox. [12]

4 Describe:
   a) the formation and circulation of lymph in a mammal. [10]
   b) the role of the mammalian lymphatic system in defence against disease. [10]

18
Nervous and hormonal communication

This is the first chapter in part IV 'Response and coordination'. The overview notes below show how the topics covered by the chapters in this part of the book are related.

Overview notes for part IV: Response and coordination
(Chapter numbers given in brackets)

Figure 18.1

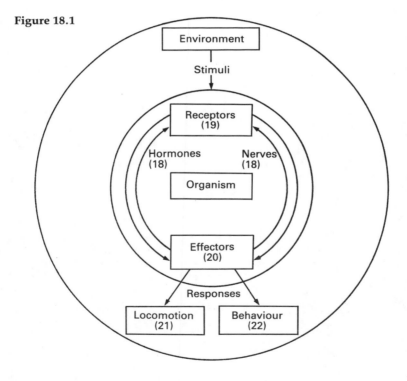

Survey and question
Survey this chapter and make overview notes in pattern form.

Read

THE NERVE CELL AND ITS IMPULSE
THE STRUCTURE OF NERVE CELLS

Recall

1 a) Copy figure 18.2 and label the structures 1–8.

Figure 18.2 *A plan of the nervous system*

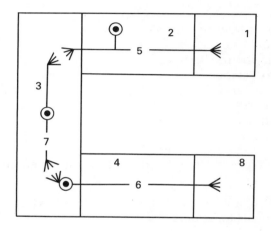

b) Indicate at 5, 6 and 7 the direction of the nerve impulse.

2 Explain the meaning of: unipolar, bipolar and multipolar neurones.

3 Make an annotated diagram of a vertebrate motor neurone.

Read

INVESTIGATING THE NERVE IMPULSE
THE ELECTRICAL NATURE OF THE NERVE IMPULSE
THE IONIC BASIS OF THE NERVE IMPULSE

Recall

1 a) Copy and label figure 18.3, overleaf. Indicate the direction in which the nerve impulse travels.
   b) Explain the events occurring at **A**.
   c) How may these events be investigated?

Figure 18.3 *The nature of the nerve impulse*

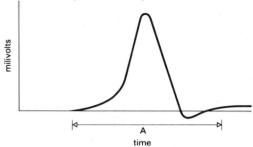

Read

PROPERTIES OF NERVES AND NERVE IMPULSES
STIMULATION
THE ALL-OR-NOTHING LAW
REFRACTORY PERIOD
TRANSMISSION SPEED
MYELIN SHEATH
AXON DIAMETER

Recall

1 List the types of stimuli that can bring about an impulse in an axon.

2 **a)** Copy figure 18.4 and complete it by drawing in the action potentials for stimuli 2 and 3.
   b) Give reasons for your answer.

Figure 18.4

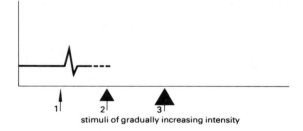

stimuli of gradually increasing intensity

3 List the factors which determine the frequency at which an axon can transmit impulses.

4 With reference to an axon, explain the meaning of the following:
   a) absolute refractory period,
   b) relative refractory period,
   c) supernormal phase.

5 Summarise diagrammatically the ionic changes that occur when a myelinated axon becomes depolarised.

6 a) Describe the differences in structure between the axons of vertebrates and invertebrates.
   b) How do these differences relate to the speed of transmission of nerve impulses?

Read
THE SYNAPSE
STRUCTURE OF THE SYNAPSE
TRANSMISSION ACROSS THE SYNAPSE
THE TRANSMITTER SUBSTANCE
THE NERVE MUSCLE JUNCTION
THE ACTION OF DRUGS AND POISONS
NORADRENALINE
SUMMATION AND FACILITATION
INHIBITION
ACCOMMODATION
THE ROLE OF SYNAPSES IN THE NERVOUS SYSTEM

Recall
1 a) Define synapse, synaptic knob and neuromuscular junction.
   b) State two differences between synaptic transmission and axon transmission.
2 Copy figure 18.12C in BAFA and annotate it to describe synaptic transmission. Include in your annotations:
   a) the nature of the nerve impulse,
   b) the role of the mitochondria,
   c) the width of the synaptic cleft,
   d) the name of the transmitter substance,
   e) the process by which transmitter substance moves across the synaptic cleft,
   f) the effect of transmitter substance on the post-synaptic membrane,
   g) the role of cholinesterase.
3 Describe how synaptic transmission can be investigated experimentally.
4 Explain the following:
   a) If shot with an arrow tipped with curare, an animal becomes rigid and unable to move.
   b) A person suffering from strychnine poisoning will undergo violent muscular convulsions.
   c) A person who has taken LSD or mescaline experiences hallucinations.

5 With reference to nerve cells, compare and contrast:
   a) spatial and temporal summation,
   b) inhibition and accommodation.

6 Make brief notes to describe the role of synapses in the functioning of the nervous system.

Read

REFLEX ACTION
STRUCTURE OF A GENERALISED REFLEX ARC
THE VERTEBRATE REFLEX ARC

Recall

1 Giving named examples, define (a) reflex action and (b) a reflex arc.

2 a) Copy figure 18.15A in BAFA and annotate it to explain the events that would occur if you placed your hand on a hot object.
   b) Your diagram shows nerve impulses passing only from the receptors to the effectors in your arm, causing you to remove it from the hot object. If you were to cry out and move away from the hot object, muscles (effectors) in the larynx and legs would have been stimulated. How would this be brought about?

Read

THE ORGANIZATION OF THE NERVOUS SYSTEM
THE VERTEBRATE CNS
THE PERIPHERAL NERVES

Recall

1 a) Copy figure 18.5 and complete it by filling in i–x.
   b) Why is this an oversimplification?

Figure 18.5 *Diagram summarising the subdivisions of the vertebrate nervous system*

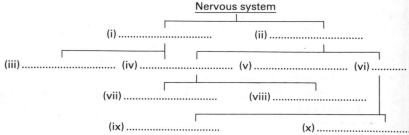

2 Copy figure 18.6 and label it to show the following:
   a) the four divisions of the brain and the spinal cord,
   b) the olfactory lobe,
   c) the pineal body,
   d) the pituitary gland,
   e) the optic lobe,
   f) the cerebrum,
   g) the cerebellum,
   h) the medulla oblongata,
   i) anterior and posterior choroid plexus,
   j) ventricles,
   k) central canal filled with cerebrospinal fluid.

Figure 18.6 *Diagram summarising the structure of the vertebrate CNS*

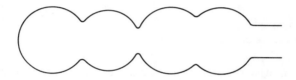

3 How is the CNS (a) protected and (b) nourished?
4 Explain the meaning of, and give named examples of:
   a) a mixed nerve,
   b) a sensory nerve,
   c) a motor nerve.
5 Why are the cranial and spinal nerves classified as voluntary?
6 In a mammal, which nerves are responsible for carrying impulses which give rise to the following: (a) hearing, (b) smell, (c) salivation, (d) sight and (e) movement of eyeballs.

Read
THE AUTONOMIC NERVOUS SYSTEM

Recall
1 a) Why is the autonomic nervous system described as involuntary?
   b) Explain why this is not a strictly correct description.

2 a) Copy and complete table 18.1

Table 18.1 *Summary of the functions of the autonomic nervous system*

Target organ	Sympathetic effect	Parasympathetic effect
Iris		
Tear gland		
Salivary gland		
Heart		
Lungs		
Stomach		
Intestine		
Pancreas		
Adrenals		
Bladder		
Genital organs		

   b) Summarise diagrammatically the general organisation of the vertebrate autonomic nervous system, adding notes about the nature of the transmitter substances involved.

Read
FUNCTIONS OF THE VERTEBRATE BRAIN IN GENERAL
FUNCTIONS OF THE MAIN PARTS OF THE BRAIN IN A FISH
FUNCTIONS OF THE MAIN PARTS OF THE BRAIN IN MAMMALS
LOCALIZATION IN THE CORTEX

Recall
1 a) State the general function of the vertebrate brain.
   b) Describe briefly the role of the ganglia in carrying out this function.
2 Make brief notes on the following methods of determining the function of particular regions of the brain: (a) physiological analysis and (b) study of behaviour patterns.
3 Copy and complete table 18.2.

Table 18.2 *Comparison of the functions of the main parts of the brain in a fish and a mammal*

	FISH		MAMMAL	
Functions	Effect of removal (if known)	Functions		Effect of removal (if known)
Forebrain		Cerebral hemispheres		
Tweenbrain		Pineal body		
		Hypothalamus		
		Pituitary gland		
Midbrain		Corpora quadrigemina		
		Red nucleus		
Hindbrain		Cerebellum		
		Medulla oblongata		

4 a) How may the functions of the cerebral cortex be investigated?
   b) Copy and complete table 18.3.

Table 18.3 *Summary of the functions of the regions of the cerebral cortex*

Example	Functions
Sensory	
Motor	
Association	

5 Outline the evidence which indicates the functions of the frontal lobes.

6 Define the following: (a) alpha waves, (b) beta waves, (c) delta waves.

7 Describe how the electrical activity of the brain may be used in the diagnosis of mental disorders.

Read
THE PRIMITIVE NERVOUS SYSTEM
INTERNEURAL AND NEUROMUSCULAR FACILITATION
THROUGH-CONDUCTION
THE ROLE OF THE BRAIN IN LOWER ANIMALS
CEPHALIZATION

Recall

1 State (a) two structural similarities and (b) two structural differences between a nerve net and a CNS.

2 a) Draw a diagram showing the structure of a nerve net.
  b) On your diagram indicate the routes that could be taken by a nerve impulse resulting from the stimulation of a sensory cell.
  c) Annotate your diagram to show the routes taken by nerve impulses when sea anemones are stimulated so that the following occur:
     (i) one tentacle pulls away from the stimulus, (ii) all the tentacles pull away from the stimulus and (iii) the whole animal closes up when stimulated.

3 a) What evidence is there to suggest that, in addition to a nerve net, sea anemones also possess through-conduction tracts?
  b) State the function of the through-conduction tracts.

4 a) Define cephalization.
  b) State the possible reasons for cephalization and the development of a brain.

5 Describe the experimental evidence for the functions of the brain in lower animals.

6 Draw a diagram summarising the structure and functions of a nervous system in an invertebrate.

Read
HORMONAL COMMUNICATION
HORMONAL COMPARED WITH NERVOUS COMMUNICATION
PRINCIPALS OF HORMONE ACTION ILLUSTRATED BY THE THYROID
   GLAND
CONTROL OF THYROXINE PRODUCTION
THE ROLE OF THE PITUITARY GLAND

NEUROSECRETION
HOW HORMONES CONTROL CELLS

Recall

1 Define (a) hormone, (b) endocrine organ and (c) target organ.
2 a) Copy and complete table 18.4.

Table 18.4 *The differences between hormonal and neural communication*

Neural	Hormonal
1. The 'message' is an all-or-nothing action potential transmitted along a nerve fibre. 2. etc.	The 'message' is a chemical carried in the bloodstream. 2. etc.

b) State one similarity between the two systems.

3 By means of an annotated diagram based on figure 18.28 in BAFA, describe
   a) how the thyroid gland fulfills the basic requirements of an endocrine organ, and
   b) the process by which dietary iodine is converted into thyroxine.

4 a) What is the function of thyroxine?
   b) Summarise diagrammatically the mechanism by which the level of thyroxine in the bloodstream is kept constant.

5 Copy and complete table 18.5.

Table 18.5 *The effects of abnormal thyroxine production in humans*

Name of condition	Symptoms	Remedy
Hypothyroidism Due to .... a Young people .... b Adults		
Hyperthyroidism Due to		

6   a) Define neurosecretion.
    b) Explain why neurosecretion is regarded as indicating the close connection between the nervous and endocrine systems.

7   Copy figure 18.32 in BAFA and annotate it to show:
    a) the receptor site for hormone molecules,
    b) the possible ways by which cyclic AMP may bring about the appropriate response to the hormone,
    c) the role of phosphodiesterase.

Review

1   a) Make a fully labelled diagram to show the structure of a mammalian neurone. [4]
    b) Explain how the nerve impulse is generated and transmitted between neurones. [10]
    c) Describe an experimental procedure for detecting nerve impulses. [6]

2   a) Make a fully labelled diagram to show the pathway of a reflex arc in a mammal. [8]
    b) Explain the transmission of nerve impulses across a synapse and a nerve muscle junction. [6]
    c) Explain the effects of the following on the nervous system: atropine, curare and strychnine. [6]

3   a) Make a fully labelled diagram to show the structure of the mammalian brain. [8]
    b) Compare the functions of the main parts of the brain in a fish and a mammal. [8]
    c) List the functions of the brain in lower animals. [4]

4   a) Make a large, fully labelled diagram to show the distribution of endocrine glands in a human. [8]
    b) List the functions of each of these glands. [4]
    c) Describe the functions of the pituitary gland in a mammal. [8]

5   a) Give a full account of the structure of the vertebrate nervous system. [15]
    b) Distinguish between the functions of the parasympathetic and sympathetic parts of the autonomic nervous system. [5]

6   Discuss the role of the nervous and hormonal control mechanisms in living organisms. [20]

19

Reception of stimuli

Survey and question
Survey this chapter and make overview notes in pattern form.

Read
RECEPTION OF STIMULI
SENSORY CELLS

Recall
1 Define (a) sensory cell, (b) receptor and (c) sense organ.
2 Copy figure 19.1 and complete it by giving named examples of each type of receptor.

Figure 19.1 *Classification of receptors*

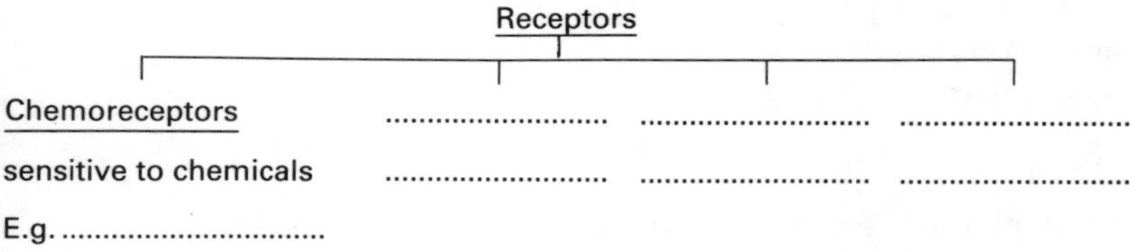

```
                          Receptors
         ┌─────────────┬─────────────┬─────────────┐
Chemoreceptors       .............  .............  .............
sensitive to chemicals .............  .............  .............
E.g. .........................
```

3 a) Make a labelled diagram to show the structure of a secondary receptor.
   b) How do primary receptors differ from secondary receptors?

Read
FUNCTIONING OF A SENSORY CELL
HOW DOES A RECEPTOR CELL WORK?
FREQUENCY OF DISCHARGE
ADAPTATION
FUSION OF STIMULI

Recall

1. **a)** State the function of a sensory cell.
   b) Describe how the functioning of receptors can be demonstrated.

2. By means of annotated diagrams, show how
   a) a generator potential is thought to develop in a mechanoreceptor.
   b) an action potential develops in the axon of a receptor.

3. If a mechanoreceptor were stimulated as described below (a–e), explain how each would affect the action potential:
   a) a very weak stimulus (i.e. below threshold)
   b) a stimulus marginally above threshold
   c) a strong, short stimulus
   d) a strong, maintained stimulus
   e) several strong, short stimuli repeated at rapid frequency.

4. **a)** Describe how adaptation is believed to be brought about.
   b) What is the importance of adaptation?

Read

INITIAL EVENTS IN THE RECEPTION OF STIMULI
MUTUAL INHIBITION
INHIBITION THROUGH EFFERENT NERVES
SENSITIVITY
PRECISION

Recall

1. **a)** Name the two types of sensory cell in the retina.
   b) State the functions of each.

2. Describe how a photoreceptor cell in a mammal is adapted to its function.

3. **a)** Describe how a photoreceptor cell can convert light energy into a nerve impulse.
   b) Explain why it is difficult to see when first entering a dimly lit room after being in bright light.

4. Using a named example, (a) define mutual inhibition and (b) explain the importance of mutual inhibition in vision.

5. Using a named example, (a) define summation and (b) explain the importance of retinal convergence in vision.

6 a) Define visual acuity.
   b) Account for the high visual acuity of the vertebrate eye.

Read
SENSE ORGANS
THE MAMMALIAN EYE
THE RETINA
COLOUR VISION
THE COMPOUND EYE

Recall
1 a) From memory, make a fully labelled diagram to show the structure of the mammalian eye.
   b) Annotate your diagram to show the function of each labelled structure.
2 Describe how the lens can accommodate for viewing both near and distant objects.
3 a) Make a fully labelled diagram to show the structure of rods and cones.
   b) Annotate your diagram to show how these photoreceptors are adapted to their particular functions.
4 Answer questions 1–5 on page 308 in BAFA.
5 a) Describe the experimental evidence which shows that there are three types of cones in the retina.
   b) How is this evidence used to explain colour vision?
6 a) Make a fully labelled diagram of an ommatidium in longitudinal section.
   b) Annotate your diagram to describe the function of each labelled structure.
7 Write brief explanations of the following:
   a) Insects are incapable of perceiving two or more closely placed objects as separate images.
   b) Insects are capable of detecting movement over a wide field.

Read
THE MAMMALIAN EAR
HEARING
BALANCE

Recall

1. a) State the two main functions of the mammalian ear.
   b) From memory, make a fully labelled diagram to show the structure of the mammalian ear.
   c) Annotate your diagram to describe the function of each labelled structure.

2. Explain the role of the tympanic and basilar membranes in the discrimination of (a) intensity (loudness) of sound and (b) pitch (high or low sounds).

3. What effect would each of the following have on the basilar membrane
   a) a quiet, low-pitched sound,
   b) a loud, high-pitched sound?

4. Describe how energy from sound waves is thought to be converted into an action potential.

5. a) From memory, make a diagram to show the structure of the entire vestibular apparatus of the mammalian ear.
   b) Annotate your diagram to describe the function of each labelled structure.

Review

1. a) Make a fully labelled diagram to show the vertebrate eye in section. [8]
   b) Briefly describe the structure and function of rods and cones. [8]
   c) Give the symptoms of colour blindness and give an explanation for its occurrence. [4]

2. a) Make a fully labelled diagram to show the structure of the mammalian ear in section. [8]
   b) Briefly describe how sound waves are transmitted to the sensory cells in the cochlea. [7]
   c) How does the ear discriminate between sounds of different intensity and pitch? [5]

3. Explain the following:
   a) The biological significance of sensory adaptation. [5]
   b) How long and short sight may be corrected. [5]
   c) How the eyes of nocturnal mammals differ from those of diurnal mammals. [5]
   d) How the mammalian ear functions as an organ of balance. [5]

4. Write an essay on photoreceptors. [20]

20

Effectors

Survey and question
Survey this chapter and make overview notes in pattern form.

Read
DIFFERENT TYPES OF MUSCLE
PROPERTIES OF SKELETAL MUSCLE
THE SINGLE SWITCH
SUMMATION
TETANUS
FATIGUE
ELECTRICAL ACTIVITY IN MUSCLE

Recall
1 a) Define an effector.
   b) Copy and complete figure 20.1.

Figure 20.1

```
                    Effectors
            ┌──────────┴──────────┐
under nervous control       independent of nervous control
E.g. 1 ......................    E.g. 1 ......................
     2 ......................         2 ......................
```

2 a) Describe how the properties of muscle can be investigated experimentally.
   b) Make a drawing to show the kymograph trace obtained from a single muscle twitch. Annotate your diagram to explain (i) the latent period, (ii) period of contraction, (iii) period of relaxation and (iv) refractory period.

3 Copy and complete table 20.1.

Table 20.1 *Different types of muscle*

	Location	Innervation	Type of contraction
Skeletal (striated)			
Visceral (smooth, non-striated)			
Cardiac (striated)			

4 Copy figure 20.2 and add notes to explain how each trace was obtained.

Figure 20.2 *Kymograph traces obtained from frog muscle*

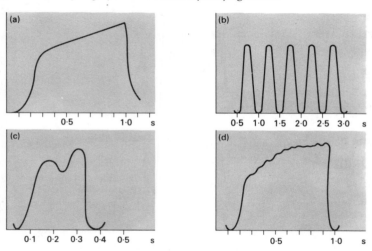

5 List those properties of muscle cells which are similar to neurones.

Read

HOW MUSCLE CONTRACTS
THE LIGHT MICROSCOPE STRUCTURE OF MUSCLE
THE FINE STRUCTURE OF MUSCLE
THE CHEMISTRY OF MUSCLE

Effectors 141

Recall
1 Make labelled drawings to show:
   a) a whole muscle,
   b) a striated muscle fibre in longitudinal section,
   c) one sarcomere as viewed with the electron microscope,
   d) one sarcomere – vertical section to show the position of the thick and thin filaments.

2 Describe the evidence from (a) electron microscopy and (b) biochemistry which has led to the interpretation of muscle structure as shown in your diagram for question 1c and d.

Read
THE SLIDING FILAMENT HYPOTHESIS
WHAT PROPELS THE FILAMENTS?
MUSCLE SPINDLES
TENDON ORGANS

Recall
1 a) By means of an annotated diagram describe Huxley and Hansons' sliding filament hypothesis.
  b) What evidence is there to support this hypothesis?

2 a) By means of an annotated diagram explain the ratchet hypothesis.
  b) Describe the evidence that supports this hypothesis.

3 Describe the evidence which indicates how electrical and mechanical processes are linked when striated muscle contracts.

4 a) Compare and contrast the properties of a muscle spindle and a tendon organ.
  b) How could you investigate the properties of these stretch receptors?

Read
OTHER EFFECTORS
CHROMATOPHORES
ELECTRIC ORGANS
LIGHT PRODUCING ORGANS
THE NEMATOBLAST

Recall

1 Copy and complete table 20.2.

Table 20.2 *Effectors*

	Location in named example(s)	Mode of action	Function
Chromatophores (pigment cells) **Electric organs** etc.			

Review

1. **a)** By means of fully labelled diagrams, compare the structure of smooth, striated and cardiac muscle as seen with the light microscope. [10]
   b) State where each type of muscle may be found in the mammal. [4]
   c) Explain how the structure of each type of muscle is related to its function. [6]

2. **a)** Describe the fine structure of striped muscle and outline the procedures that have been used to determine the biochemistry of its components. [12]
   b) Give a full explanation of the mechanism of contraction of striped muscle. [8]

3. Survey the range of effectors in animals. [20]

4. The gastrocnemius muscle of a frog was removed together with its nerve supply. The nerve–muscle preparation was set up in an apparatus which recorded contractions of the muscle when the nerve was stimulated. The following traces were obtained (figure 20.3). Points of stimulation are recorded below each trace. One hertz (Hz) is equivalent to one stimulation per second.
   a) Examine the five traces.
      (i) Compare traces **A** and **B** and account for the differences between them.
      (ii) The shortest interval between stimuli is the same in **B** and **C**. Explain why the traces are different.

Figure 20.3

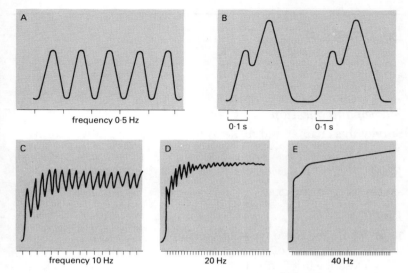

(iii) Compare traces **C, D** and **E** and account for the differences between them. [12]
b) What would happen to the muscle response in **E** if stimulation continued at the same frequency?
Give reasons for your answer. [4]
c) The gastrocnemius muscle is striated.
  (i) Write an illustrated account of the detailed structure of a striated muscle fibril as revealed by an electron microscope.
  (ii) Explain why the light bands of striated muscle become narrower as the muscle contracts. [14]   (AEB)

21

Locomotion

Survey and question
Survey this chapter and make overview notes in pattern form.

Read
SKELETON AND MUSCLES

Recall
1. a) Name the three types of skeleton.
   b) State the role of the skeleton in locomotion.
   c) Describe briefly the role of each of the three types of skeleton in locomotion.
2. List three factors to be taken into account when considering locomotion in animals.

Read
MOVEMENT IN WATER
PROPULSION IN FISHES
SUPPORT IN FISHES
BUOYANCY IN BONY FISHES
STABILITY IN FISHES

Recall
1. State one advantage and one disadvantage of water as compared to air as a medium in which locomotion occurs.
2. a) What is meant by propulsion?
   b) By means of annotated diagrams, describe the role of the tail in propulsion in a dogfish.
3. Giving named examples, describe the following types of locomotion:
   a) carangiform,
   b) anguilliform,
   c) ostraciform.

4 Describe how the following provide support and/or stability in a dogfish:
   a) pectoral and pelvic fins,
   b) the heterocercal caudal fin,
   c) anterior and posterior dorsal fins,
   d) the massiveness of the head,
   e) the streamlined shape of the body.

5 Describe the role of the following in support and/or stability in a bony fish:
   a) pectoral fins,
   b) the swim bladder,
   c) the streamlined shape of the body.

Read
THE ACTION OF CILIA AND FLAGELLA

Recall
1 By means of annotated diagrams compare and contrast the propulsive action of cilia and flagella.

Read
MOVEMENT ON LAND
THE MUSCULO–SKELETAL BASIS OF LOCOMOTION IN TETRAPODS

Recall
1 a) How does air differ from water as a medium through which to move?
   b) In note form, describe how support and propulsion are achieved in land-dwelling tetrapods.

2 a) Give named examples of two kinds of joints found in tetrapods.
   b) Make a fully labelled diagram of a ball and socket joint.
   c) Annotate your diagram to explain the function of each labelled structure and show how the limb bone is adapted to withstand compression, tension and shearing forces.

3 a) Describe how the action of muscles which move the limb bones can be investigated.
   b) On what basis are limb muscles classified?
   c) List these main groups of muscles under two headings to show which sets are antagonistic to each other.

146 *Study Guide*

4 Figure 21.1 shows very simplified diagrams of the hind leg of a tetrapod in a number of different positions.
   a) Copy them and name the type of movement each diagram illustrates.
   b) On each diagram, draw and label the muscles that are responsible for each type of action. (The abductor muscle has been drawn in for you.)

Figure 21.1

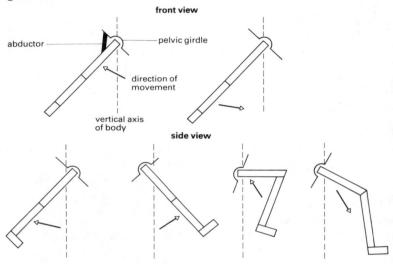

c) Describe the type of movement that is not illustrated.

Read
PROPULSION IN TETRAPODS
SUPPORT IN TETRAPODS
STABILITY IN TETRAPODS

Recall

1 a) Make a labelled diagram to illustrate the action of the hind limb in a tetrapod in propelling the body forward.
   b) Annotate your diagram to show how contraction of the retractors and extensors enables the limb to act as a lever.

2 a) By means of simple annotated diagrams, explain how the forelimb of a mammal acts as a more efficient strut than that of a reptile.

b) Explain why, in general, locomotion in mammals is more efficient than in reptiles.

3 a) State the role of the vertebral column in support in tetrapods.
   b) Make a labelled diagram of a vertebra.
   c) Annotate your diagram to explain the function of each labelled structure.

4 Figure 21.2 is a highly simplified diagram of a stationary tetrapod. Copy it, and draw four similar diagrams to show the sequence of limb movements in the locomotion of a primitive tetrapod. Indicate the centre of gravity on each diagram.

Figure 21.2 *How stability is maintained in tetrapod locomotion*

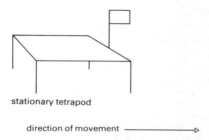

5 How is stability maintained during locomotion in the following:
   a) a horse,
   b) a rabbit,
   c) a kangaroo.

Read
MOVEMENT IN AIR
FLIGHT IN BIRDS

Recall

1 List the groups of animals which are capable of active flight.

2 Draw a diagram to show the flow of air over an aerofoil and the resultant forces.

3 Explain the following:
   a) A gliding gull does not drop like a stone in still air.
   b) Some birds can glide without losing height.

c) Gulls are able to gain height whilst gliding.
d) When there are no air currents, birds can gain height by active flight.

4 Explain the significance of the following:
  a) The depressor and levator muscles of a bird are rich in myoglobin.
  b) The sternum of a bird has a deep 'keel'.
  c) Many bones in the bird's skeleton are hollow.
  d) The feathers of birds increase the surface area of the forelimb.

Read

FLIGHT IN INSECTS

Recall

1 Copy figure 21.3 and complete it by labelling all structures shown. Annotate your diagrams to explain the action of the indirect flight muscles in flight.

Figure 21.3 *Action of indirect flight muscles in an insect (e.g. bee)*

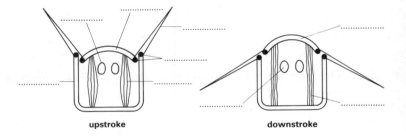

upstroke   downstroke

2 What reasons are there to suggest that the rate of wingbeat in butterflies is regulated entirely by nerve impulses, whilst in some other insects (e.g. houseflies) this is obviously not the case?

3 Explain how insects are adapted for flight in
  a) the mechanics of the musculo–skeletal system,
  b) the organisation of the nervous system,
  c) the physiology of insect muscle.

Read

AMOEBOID MOVEMENT

Recall

1. **a)** List those cells which exhibit amoeboid movement.
   b) Make an annotated diagram to illustrate amoeboid movement.

2. **a)** Briefly summarise the three theories of amoeboid movement.
   b) Describe the evidence which lends weight to one of these theories rather than the other two.

Review

1. **a)** Distinguish between transport and locomotion. [4]
   b) By means of annotated diagrams, compare the structure of xylem and bone. [10]
   c) Relate the structures of xylem and bone to their functions. [6]

2. **a)** Give an account of the tissues in the hind limb of a named mammal. [10]
   b) Explain the role of each of these tissues in locomotion. [10]

3. Describe locomotion in the following:
   a) *Ameoba*, [4]
   b) *Paramecium*, [4]
   c) a dogfish, [6]
   d) a bird. [6]

4. Compare and contrast the structure and functioning of exoskeletons and endoskeletons. [20]

5. Explain the meaning of movement and discuss its importance to living organisms. [20]

6. Consider the following data on insect flight.

	Mass in mg	Wing area mm^2	Number of wing beats per second
Housefly	12	24	200
Lacewing	25	125	52
Greenfly	2	8	100
Caddis fly	16	68	90
Noctuid moth	160	240	225

For each type of insect:
a) Calculate the ratio of wing area (mm^2) to mass (mg) for each insect. [4]

b) Construct a graph to show the relationship between this ratio and the number of wing beats per second. [6]
c) What relationship exists between the ratio of wing area to mass and the number of wing beats per second? [2]
d) The above group of insects all fly at about five miles per hour. On your graph plot the position of a moth with a wing area to body mass ratio of 4.5 to 1, and a wing beat of 150. Would such a moth fly faster than or slower than the group of insects previously plotted? Explain your answer. [4]
e) How would you have measured the number of insect wing beats per second? [2]
f) How would you have calculated the wing area of the insects? [2]
g) What features of insect structure and physiology have contributed to their success as flying animals? [5]   (AEB)

22

Behaviour

Survey and question
Survey this chapter and make overview notes in pattern form.

Read
A WORD ABOUT TECHNIQUES
TYPES OF BEHAVIOUR

Recall
1 Define (a) behaviour and (b) ethology.

2 Explain the necessity for the following in ethology:
   a) activity recorders, and
   b) maintaining the animal in as natural conditions as possible.

3 a) List some of the recording techniques used in ethology.
   b) What is the ultimate aim of ethologists?
   c) Why is it not always possible to achieve this aim?

4 Using named examples explain the meaning of:
   a) species-characteristic behaviour,
   b) individual-characteristic behaviour.

Read
REFLEX ACTION
ESCAPE RESPONSES
ANALYSIS OF THE EARTHWORM'S ESCAPE RESPONSE
ESCAPE RESPONSE OF THE SQUID

1 a) Define reflex action.
   b) Give an example of a reflex action.

2 Describe the escape responses of an earthworm and a squid.

3 Explain how the neural basis of the earthworm's escape response may be investigated.

4 What role do each of the following play in the escape response of the earthworm:
   a) the median giant axon,
   b) the lateral giant axons?

5 What role do each of the following play in the escape response of the squid:
   a) the stellate ganglia,
   b) the giant axons?

Read
ORIENTATION
TAXIS
KINESIS

Recall
1 a) State the two main types of orientation response.
   b) Give two named examples of orientation responses.
   c) State two ways in which an orientation response differs from a reflex action.

2 Figure 22.1 shows a choice chamber indicating the position of woodlice which have been placed in the dry side of the chamber.

Figure 22.1

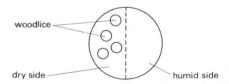

   a) Make a copy of this diagram and construct a similar one to show the likely positions of the woodlice five minutes later.
   b) Name the type of orientation behaviour exhibited by the woodlice in this situation.

3 How would you investigate:
   a) chemotaxis in a flatworm,
   b) phototaxis in *Euglena*?

4 State one similarity and one difference between a taxis and a kinesis.

5 Figure 22.2 shows a petri dish which previously contained two flatworms. The dish has been treated with charcoal as described

Figure 22.2 *Petri dish showing tracks of flatworms*

in BAFA to show the trails made by the flatworms. Make a copy of the diagram and annotate it to explain the shape of the tracks left by the flatworms.

Read
ANALYSIS OF SPECIES-CHARACTERISTIC BEHAVIOUR
THE ROLE OF STIMULI
RELEASERS
STIMULUS SELECTION
THE FUNCTION OF RELEASERS

Recall

1 a) State the factors that control species-characteristic behaviour.
   b) Describe briefly how one might study the effects of these factors on behaviour.
   c) Using a named example, explain what the results of such studies indicate about species-specific behaviour.

2 a) Explain how the popular meaning of instinct differs from that used by ethologists.
   b) Using a named example, explain the meaning of motivation.

3 a) List the three kinds of stimuli.
   b) Give an example of each kind.

4 For each of the following examples, identify the type of stimuli involved.
   a) Woodlice were kept for 24 hours in a dry environment. They were then placed in the dry half of a choice chamber, and moved about actively. Within ten minutes, the woodlice had crossed to the humid side and had become stationary.
   b) If placed in bright illumination, blowfly larvae will crawl away from the light. They become stationary on reaching a dark crevice.

c) The cuckoo lays its eggs in the nests of a number of different species of birds, particularly reed warblers. Adult reed warblers will feed a young cuckoo which hatches in their nest, in spite of its large size. This is because the cuckoo chick gapes its beak, exposing the red colour of the beak lining, whenever the adults approach the nest. This action on the part of the cuckoo chick elicits feeding behaviour in the adults.
d) In (c), what evidence is there that stimulus selection is involved?

5 Using named examples, explain the importance of:
   a) threat displays,
   b) appeasement displays.

6 Give a named example of:
   a) an internal motivational stimulus,
   b) an internal terminating stimulus.

Read
SYNCHRONIZATION IN SEXUAL BEHAVIOUR
MOTIVATION
PHEROMONES
THE ROLE OF HORMONES
THE HYPOTHALAMUS AND BEHAVIOUR
DISPLACEMENT ACTIVITY
VACUUM ACTIVITY

Recall

1 Table 22.1 indicates some aspects of behaviour in the male stickleback. Copy the table and complete it by writing brief notes on the stimuli involved.

Table 22.1 *A summary of the sexual behaviour of the male three-spined stickleback*

Behaviour	Stimuli involved
Migratory	
Territorial	
Nest building	
Courtship	
Mating	
Fanning	

2 a) What is a pheromone?
   b) Using two named examples, describe the effect of pheromones.
3 Summarise briefly the possible effects of hormones on (a) the brain and (b) receptors and effectors.
4 Giving a named example for each, define (a) displacement activity and (b) vacuum activity.

Read
LEARNING
HABITUATION
ASSOCIATIVE LEARNING
THE CONDITIONED REFLEX
TRIAL AND ERROR LEARNING

Recall
1 a) What is meant by learning?
   b) How does learned behaviour differ from instinctive behaviour?
   c) Why does learned behaviour differ from one individual to another?
2 List five categories of learning.
3 Using the earthworm as an example, explain (a) the meaning of habituation and (b) the neurological basis of habituation.
4 a) How does habituation differ from sensory adaptation?
   b) What is the value of habituation to an animal?
5 a) What is meant by associative learning?
   b) Figure 22.3 is a summary of Pavlov's experiment on conditioned reflexes. Complete it by indicating the nature of the stimuli (S) and responses (R).

Figure 22.3

S_1 — — — — — — — — — — — — — — — → R (unconditioned reflex)
(..............................)   (..............................)

$S_1 + S_2$ — — — — — — — — — — — — — → R
(.......................... + ...)   (..............................)

S_2 — — — — — — — — — — — — — — — → R (conditioned reflex)
(..............................)   (..............................)

c) Of what value is associative learning to animals in the wild?
6 Examine figure 22.17 in BAFA and answer the following questions.
  a) Which group of animals did not receive a reward?
  b) Which group received regular rewards?
  c) How does trial and error learning differ from conditioning?
7 a) What evidence is there to indicate that earthworms and octopuses are capable of learning?
  b) Explain why an octopus is able to learn more quickly by trial and error than an ant.
8 Using the results shown in figure 22.19 in BAFA, say which of the following statements is true, giving reasons for your answers.
  a) Learning does not occur unless rewards are given.
  b) Ants go through the maze the first time by conditioned reflex and subsequently by trial and error.
9 In figure 22.19 in BAFA:
  a) Why is it necessary for the foodbox at the end of the maze to be empty?
  b) What evidence is there that learning has occurred?

Read
LEARNING AND THE OCTOPUS BRAIN
A NEURAL THEORY OF LEARNING
A BIOCHEMICAL THEORY OF LEARNING
EXPLORATORY LEARNING
IMPRINTING
INSIGHT LEARNING

Recall
1 a) What is meant by short term and long term memory?
  b) What evidence is there to indicate the existence of these two memory systems?
2 What is the main difference between a neural theory of learning and a biochemical theory of learning?
3 Of what value are the following in the lives of animals (a) exploratory learning, and (b) imprinting?
4 a) Why is insight learning regarded as being the highest form of learning?
  b) Describe one experiment to investigate insight learning.

Review

1. **a)** Distinguish between:
   - (i) instinctive and learned behaviour,
   - (ii) habituation and adaptation. [8]

   b) Using named examples, explain the meaning of the following:
   - (i) displacement activity,
   - (ii) imprinting,
   - (iii) trial and error learning. [12]

2. **a)** Distinguish between a taxis and a kinesis. [4]

   b) Describe, with full experimental details, how you could investigate:
   - (i) kinesis in woodlice,
   - (ii) chemotaxis in flatworms. [16]

3. **a)** By means of an annotated diagram, explain the events that would occur in the nervous system if someone's hand was subjected to a mild electric shock. [12]

   b) Describe how this reflex could become conditioned. [8]

4. Using named examples, explain the biological significance of the following:
   - **a)** territorial behaviour, [7]
   - **b)** courtship, [7]
   - **c)** parental care. [16]

5. Write an essay on social insects. [20]

23

Cell division

This is the first chapter in part V 'Reproduction, development and heredity'. The overview notes below show how the topics covered by the chapters in this part of the book are related.

Overview notes for part V: Reproduction, development and heredity
(Chapter numbers given in brackets.)

Figure 23.1

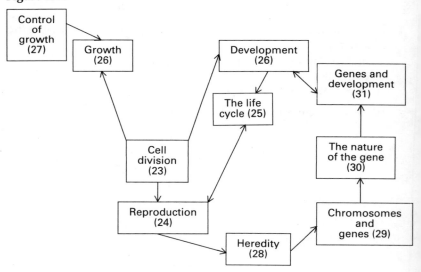

Survey and question
Survey this chapter and make overview notes in pattern form.

Read
MITOSIS

Recall

1 Figure 23.2 shows the main stages in the mitotic division of a cell containing two chromosomes.
   a) Copy the diagrams, identify each stage illustrated, arrange them in the correct sequence and label and annotate each diagram to explain what is happening.
   b) Does the cell illustrated in the diagrams come from an animal, a lower plant or a higher plant? Give reasons for your answer.

Figure 23.2

2 As a result of mitosis, how do the daughter cells compare to the parent cell in terms of: (a) chromosome number, and (b) genetic constitution of the chromosomes?

3 List two processes which involve mitosis.

4 Where would you expect to find mitosis taking place in (a) an animal, and (b) a plant?

Read

MEIOSIS

Recall

1 Figure 23.3 summarises the process of reproduction in humans. Copy and complete it by filling in the gaps using words from the following list (words may be used more than once): fertilization, zygote, haploid, gametes, meiosis, diploid, mitosis.

Figure 23.3

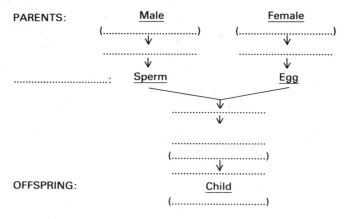

2 Figure 23.4 shows the main stages in the meiotic division of a cell. Make copies of them, identify the stage illustrated, arrange them in the correct sequence and label and annotate each diagram to explain what is happening.

3 As a result of meiosis, how do the daughter cells compare with the parent cell in terms of (a) chromosome number, and (b) genetic constitution of the chromosomes?

4 List two processes which involve meiosis.

5 Where would you expect to find meiosis occurring in (a) an animal, and (b) a plant?

6 Construct a table to compare and contrast the following aspects of mitosis and meiosis: pairing of homologous chromosomes, number of divisions, number of daughter cells produced, genetic similarity of parent and daughter cells, number of chromosomes in parent and daughter cells.

Figure 23.4

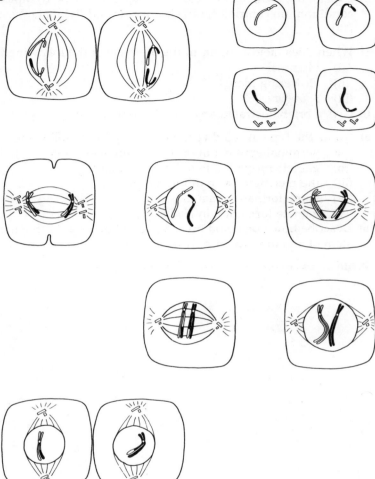

Review

1. a) Give an illustrated account of mitosis in an animal cell. [10]
   b) Explain how mitosis in an animal cell differs from that in a plant cell. [5]
   c) What happens to the cell organelles during mitosis? [3]
   d) What is the biological significance of mitosis? [2]

2. a) Outline briefly the essential differences between mitosis and meiosis. [6]

b) Explain the meaning and biological significance of: (i) diploid condition, (ii) centrioles, (iii) chromatids and (iv) metaphase. [14]

3 a) When does meiosis occur in the life cycle of the following?
    (i) Mammal.
    (ii) Flowering plant.
    (iii) *Spirogyra*. [6]
b) Give a brief illustrated account of meiosis. [14]

4 a) Name the type of cell division occurring in the following:
    (i) germinal epithelium of the mammalian testis.
    (ii) gamete formation in a flowering plant.
    (iii) the meristem in an onion root tip.
    (iv) spore formation in a moss.
    (v) gamete formation in *Hydra*. [5]
b) Describe how you could investigate cell division in the meristem of an onion root tip. [15]

5 Write an essay on chromosome numbers. [20]

24

Reproduction

Survey and question
Survey this chapter and make overview notes in pattern form.

Read
THE GENETIC IMPORTANCE OF SEX: BACTERIAL CONJUGATION
EVOLUTIONARY DEVELOPMENT OF GAMETES

Recall
1 Summarise by means of a flow diagram or brief notes, Lederberg and Tatum's experiment. What is the significance of the experiment?

2 Write brief definitions of the following:
   a) syngamy,
   b) heterogametes,
   c) isogametes,
   d) anisogametes.

Read
THE SPERMATOZOON
THE EGG CELL
FERTILIZATION
EVOLUTION OF REPRODUCTIVE METHODS

Recall
1 Make large, fully labelled diagrams of a human sperm and a human egg. Indicate the size of each on the diagrams.

2 The notes below are a summary of the main events in the process of fertilization. Copy and complete them by filling in the gaps.
   a Sperm penetrates jelly coat surrounding egg.
   b Sperm comes into contact with ....
   .... and the ....
   takes place.

c .... becomes thickened and lifts off from the plasma membrane – now known as the .... Prevents entry of further sperms.
   d .... becomes detached from ....
   e Sperm nucleus fuses with egg nucleus, restoring the .... Spindle forms.
   f Zygote undergoes its first .... division.

3 List two advantages of internal fertilization over external fertilization.

Read

SEXUAL REPRODUCTION IN MAN
GAMETOGENESIS
HISTOLOGY OF THE TESTIS
HISTOLOGY OF THE OVARY
WHAT BRINGS SPERM AND EGGS TOGETHER?
DEVELOPMENT OF THE ZYGOTE

Recall

1 How does the growth phase of gametogenesis differ in the two sexes?

2 How does the maturation phase of gametogenesis differ in the two sexes? (Two differences).

3 Which of the following are haploid: spermatid, primary oöcyte, secondary spermatocyte, oögonia, first polar body, primordial germ cell?

4 a) Below are the names of some of the stages in oogenesis and spermatogenesis. Copy them in the correct sequence and add notes to describe how gametogenesis occurs: primary oöcyte, spermatozoa, primordial germ cell, first polar body, oögonia, secondary spermatocyte, secondary oöcyte, spermatids, additional polar bodies, primary spermatocyte, ovum, spermatogonia.
   b) In which parts of the gonads do oögonia and spermatogonia develop?

5 Summarise diagrammatically the development of:
   a) a Graafian follicle,
   b) the corpus luteum.

6 From memory, make fully labelled diagrams to show the structure of the human male and female reproductive systems.

7 Write brief notes on each of the following in humans:
   a) copulation,
   b) fertilization,
   c) implantation.

8 Describe how the placenta carries out each of the following functions:
   a) protection of the foetus,
   b) providing the foetus with nourishment,
   c) as a respiratory organ,
   d) as an excretory organ.

9 a) Make a fully labelled diagram to show the structure of a chorionic villus, and its relationship with the foetal and maternal circulations.
   b) Explain how foetal blood is able to absorb oxygen at the low partial pressure found in the placental circulation.

10 Define:
   a) viviparity,
   b) parturition,
   c) gestation period,
   d) lactation,
   e) puberty,
   f) menopause.

Read

THE SEXUAL CYCLE
IN THE EVENT OF PREGNANCY
THE SEXUAL CYCLE IN OTHER MAMMALS
SEXUAL ACTIVITY IN THE HUMAN MALE

Recall

1 What is meant by:
   a) sexual cycle, and (b) menstrual cycle?

2 Describe the changes that occur in: (a) the endometrium, and (b) a Graafian follicle, during one menstrual cycle.

3 a) List the hormones which are responsible for developments in: (i) the uterine wall and (ii) the Graafian follicle, during the menstrual cycle.
   b) Where is each of these hormones produced?

4 a) Copy the sequence of hormones in figure 24.1 and annotate it to show how the production of hormones is controlled in the menstrual cycle.

Figure 24.1 *Hormone production in the menstrual cycle*

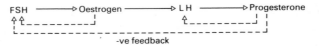

b) Briefly describe the functions of each of the above hormones in the menstrual cycle.

5 a) Draw a simple graph based on figure 24.2 to show the relative levels of oestrogen and progesterone during a single menstrual cycle.

Figure 24.2

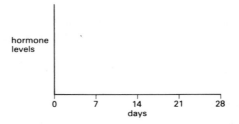

b) Below the graph, indicate the changes that occur in the endometrium during one menstrual cycle.

6 a) Copy the hormonal sequence in figure 24.3 and annotate it to indicate how high levels of progesterone are maintained during pregnancy.

Figure 24.3

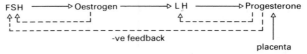

b) List the functions of progesterone during pregnancy.

7 Towards the end of pregnancy, progesterone levels begin to fall. What effect does this have on (a) the mammary glands, and (b) the uterine wall?

8 a) What is the source of progesterone during pregnancy?
   b) What is the role of oestrogen in pregnancy, parturition and lactation?

c) Name one other hormone that plays a role in parturition, and describe its function.

9 a) What is meant by breeding season?
   b) List some of the factors which may regulate breeding seasons.

10 a) List the hormones produced in human males.
   b) Identify the source of each hormone.
   c) State the functions of each hormone.

Read

SEXUAL REPRODUCTION IN THE FLOWERING PLANT
STRUCTURE OF THE FLOWER
INSIDE THE STAMEN
INSIDE THE CARPEL
TRANSFER OF POLLEN
GROWTH OF THE POLLEN TUBE AND FERTILIZATION

Recall

1 From memory, make a fully labelled diagram of a longitudinal section through a generalised actinomorphic flower.

2 a) From memory, make a fully labelled diagram to show the structure of an anther in cross section.
   b) Indicate how dehiscence occurs in this anther.

3 From memory, make labelled diagrams to show how a pollen grain develops. Indicate clearly in your diagram the stage at which meiosis occurs.

4 From memory, make a fully labelled diagram to show:
   a) a mature carpel in longitudinal section,
   b) the development of the embryo sac. Indicate clearly on your diagram the stage at which meiosis occurs.

5 a) What is meant by pollination?
   b) Copy and complete figure 24.4.

Figure 24.4

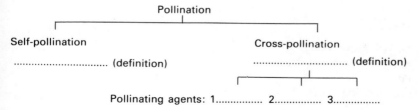

6 From memory, make fully labelled diagrams to show:
   a) germination of the pollen tube,
   b) growth of the pollen tube,
   c) fertilization.

7 a) What is meant by double fertilization?
   b) What is the significance of double fertilization?

Read

AFTER FERTILIZATION
THE FLOWERING PLANT COMPARED WITH OTHER ORGANISMS
SELF VERSUS CROSS FERTILIZATION

Recall

1 a) From memory, make a fully labelled diagram to show a longitudinal section of a fruit containing one endospermic seed.
   b) On your diagram, indicate the origin of the embryo, the endosperm, the seed coat, the fruit.

2 List two similarities and two differences between reproduction in a flowering plant and a mammal.

3 a) Define (i) dioecious and (ii) monoecious.
   b) What are the advantages and disadvantages of each?
   c) Using named examples, describe some of the mechanisms which prevent self fertilization in animals and plants.

4 Copy and complete table 24.1.

Table 24.1 *A summary of the characteristics of insect and wind pollinated flowers*

Characteristic	Insect	Wind
Petals	Brightly coloured, scented etc.	etc.

Read

PARTHENOGENESIS
ASEXUAL REPRODUCTION
FISSION
SPORE FORMATION

BUDDING
FRAGMENTATION
VEGETATIVE PROPAGATION
DISPERSAL

Recall

1. Using named examples, explain the meaning of: (a) diploid parthenogenesis, and (b) haploid parthenogenesis.

2. Copy and complete figure 24.5.

Figure 24.5

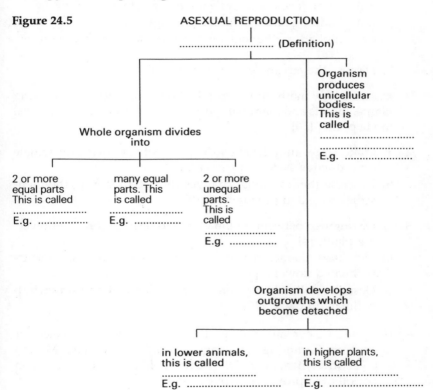

3. a) What is meant by perennating organ?
   b) What part do perennating organs play in the life cycle of flowering plants?

4. a) Define dispersal.
   b) List some of the agents of dispersal.
   c) Using named examples, list some of the forms in which animals may be dispersed.

Review

1. **a)** Compare and contrast the essential features of asexual and sexual reproduction. [2]
   b) Explain briefly how fertilization occurs in *Spirogyra*, *Fucus*, a flea, a conifer, a fish and an amphibian. [18]

2. **a)** Using at least one named organism for each method, distinguish clearly between the following methods of reproduction:
   (i) binary fission and fragmentation.
   (ii) spore formation and gamete formation.
   (iii) budding and vegetative propagation.
   (iv) diploid parthenogenesis and haploid parthenogenesis. [16]
   b) Discuss the advantages of each method. [4]

3. Describe how methods of reproduction in herbaceous flowering plants, insects and mammals appear to be adapted to a terrestrial mode of life. [20]

4. **a)** Make large, fully labelled diagrams of the male and female reproductive systems in a mammal. [10]
   b) Describe the hormonal changes that occur during the menstrual cycle and pregnancy. [10]

5. **a)** Distinguish between pollination and fertilization in a flowering plant. [4]
   b) List the characteristics of wind-pollinated and insect-pollinated flowers. [6]
   c) Describe pollination in one insect-pollinated and one wind-pollinated flower. [10]

6. **a)** Table 24.2 contains data on the breeding season of Leicester ewes in Cambridge, England (latitude 52°N), and Merino ewes in Kenya near the equator. Sheep have a gestation period of approximately five months.
   (i) Construct a graph using these data. [4]
   (ii) Suggest an explanation for the differences in the shapes of the curves for the two groups of sheep. [4]
   (iii) Briefly describe an experiment which you would carry out to test the validity of the explanation given in (ii). [3]

Table 24.2

	Percentage of flock in oestrous	
	Merino (Kenya)	Leicester (England)
January	80	58
February	76	3
March	60	0
April	64	0
May	75	0
June	80	0
July	73	0
August	60	14
September	76	78
October	78	98
November	88	85
December	90	65

b) Figure 24.6 shows the hormones present and their sequence of production in an adult human female during a normal 28-day oestrous cycle.

Figure 24.6

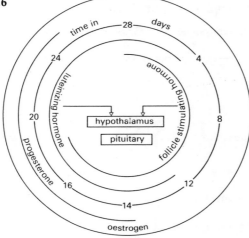

Using the information in the diagram and the knowledge you already have, describe how **all** of the hormones named on the diagram interact to control the oestrous cycle. [9]   (AEB)

25
The life cycle

Survey and question
Survey this chapter and make overview notes in pattern form.

Read
THE LIFE CYCLE OF MAN
ALTERNATION OF GENERATIONS
THE LIFE CYCLES OF MOSS AND FERN

Recall
1 Draw diagrams to summarise:
   a) the life cycle of a human,
   b) the life cycle of a plant.
On your diagrams, indicate which phases are haploid and which are diploid.

2 a) Make two copies of figure 25.1.

Figure 25.1 *Diagram summarising the life cycle of a moss (or fern)*

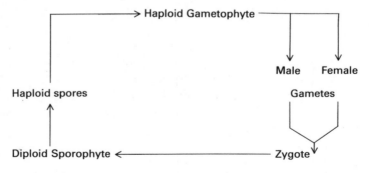

   b) Annotate one diagram to illustrate the structures involved in the life cycle of a moss and the other to illustrate the life cycle of a fern.
   c) In your diagrams, indicate where meiosis, mitosis and syngamy occur.

3 Compare and contrast the following in a moss and a fern:
   a) transference of sperm from antheridium to archegonium,
   b) dispersal of spores.

Read
REDUCTION OF THE SPOROPHYTE OR GAMETOPHYTE
CLUB MOSSES AND HORSETAILS

Recall
1 Construct a diagram similar to that shown in question 2 above to summarise the life cycles of: (a) *Spirogyra* and (b) *Selaginella*.

Read
FLOWERING PLANTS AND CONIFERS
ALTERNATION OF GENERATIONS IN ANIMALS

Recall
1 Describe the structure of the following in flowering plants: (a) sporophyte, (b) spores, (c) gametophyte, and (d) gametes.

2 Account for the reduction of the gametophyte generation in the life cycle of flowering plants.

3 a) Figure 25.2 summarises the life cycle of *Obelia*. Explain why this life cycle does not show alteration of generations.

Figure 25.2 *Diagram summarising the life cycle of* Obelia

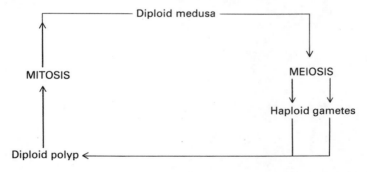

b) Draw a diagram of a life cycle which does show alternation of generations.

Review

1. a) What is meant by alternation of generations? [4]
   b) Compare and contrast the life cycles of *Hydra, Mucor*, a moss and a liverwort. [16]

2. With reference to the life cycles of a moss, *Selaginella* and a flowering plant, discuss the significance of the following: (a) the sporophyte, (b) the gametophyte, and (c) heterospory. [20]

3. a) Describe the external structure and life cycle of a named fern. [10]
   b) List those features of ferns which are considered to be adaptations to life on land. [4]
   c) Compare the life cycle of an amphibian with that of a fern. [6]

4. Write an essay on the life cycles of animals. [20]

26
Patterns of growth and development

Survey and question
Survey this chapter and make overview notes in pattern form.

Read
GROWTH
MEASURING GROWTH
THE GROWTH CURVE
RATE OF GROWTH
PERCENTAGE GROWTH
INTERMITTENT GROWTH IN ARTHROPODS

Recall
1 a) Define growth.
   b) Copy figure 26.1 and annotate it to show how each of the processes (i)–(iii) contribute to growth.

Figure 26.1 *Diagram to summarise the processes involved in growth*

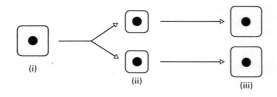

2 a) List the parameters that can be used to measure growth in an organism.
   b) For each parameter listed, describe the limitations involved.
3 What is meant by allometric growth?
4 Table 26.1 gives some figures showing the average increase in

mass with age for girls. Using these figures construct the following graphs:
a) growth curve,
b) growth rate curve,
c) percentage growth curve.

Table 26.1

Age (years)	Mass (kg)
Birth	3.4
1	9.7
2	12.3
4	16.4
6	21.0
8	26.4
10	32.0
12	39.7
14	49.0
16	53.0
18	54.4
20	54.4

5 a) Draw a growth curve for an arthropod.
 b) Explain the shape of this curve.

Read
ANIMAL DEVELOPMENT
CLEAVAGE
GASTRULATION
FORMATION OF THE NEURAL TUBE AND NOTOCHORD
FORMATION OF MESODERM
ORIGIN OF THE COELOM
FURTHER DEVELOPMENT OF MESODERM AND COELOM

Recall
1 a) State the differences between growth and development.

b) Define embryology.
c) List the three main processes which occur in the development of the embryo.
2 By means of annotated diagrams, summarise the following in a typical chordate (e.g. *Amphioxus*):
a) cleavage,
b) gastrulation,
c) neurulation,
d) development of the mesoderm.

Read
DEVELOPMENT OF BIRDS AND MAMMALS: THE EXTRA-EMBRYONIC MEMBRANES

Recall
1 Describe two ways in which development in higher chordates differs from that of *Amphioxus*.
2 a) List the extra-embryonic membranes of birds and mammals.
 b) By means of annotated diagrams, show the relationship of these embryonic membranes to a developing embryo.
 c) Annotate your diagram to describe the function of each membrane.
3 Describe the main differences between foetal and adult circulation in a mammal.

Read
LARVAL FORMS
METAMORPHOSIS

Recall
1 Using named examples, explain the meaning of: (a) larva, and (b) metamorphosis.
2 How does insect metamorphosis differ from that of amphibians?
3 Explain the biological significance of larval forms to the following animals:
a) sea anemones,
b) liver flukes,
c) the cabbage white butterfly,
d) mayflies.

4 a) Make copies of figure 26.2.

Figure 26.2 *Insect life cycles*

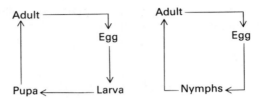

b) Indicate the type of life cycle shown by each diagram.
c) Annotate each diagram so that it describes the life cycle of a named insect.

Read

DEVELOPMENT IN THE FLOWERING PLANT
TYPES OF GERMINATION
THE PHYSIOLOGY OF GERMINATION
GROWTH AND DEVELOPMENT OF THE SHOOT AND ROOT

Recall

1 a) Draw a diagram to show the structure of a typical seed.
b) By means of simple annotated diagrams, contrast germination in a broad bean and a sunflower seed.

2 a) List the conditions necessary for germination.
b) Explain the importance of water for germinating seeds.

3 a) Name the regions in plants where growth and development occur.
b) State the differences between growth and development in plants and growth and development in animals.

4 a) Summarise diagrammatically the stages of growth and development of the tissues in the root of a flowering plant.
b) By means of simple diagrams show the changes that occur as a cell passes from the zone of cell division to the zone of elongation.

5 Describe the processes involved in the differentiation of:
a) a parenchyma cell,
b) a collenchyma cell,
c) xylem vessels,
d) sclerenchyma fibres.

Read
SECONDARY GROWTH

Recall
1. **a)** State the differences between apical growth and secondary growth.
   b) Name the two meristems involved in secondary growth.
2. **a)** Figure 26.3 represents stages in secondary growth in a dicotyledonous stem. Make copies of them and arrange them in the correct sequence.
   b) Label the tissues shown and annotate your diagrams to describe secondary growth and the development of annual rings.

Figure 26.3

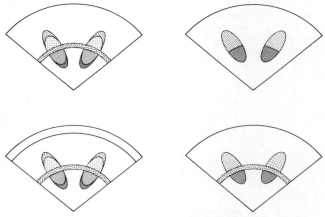

3. Describe the formation and function of lenticels.

Review
1. Compare and contrast:
   a) the structure of the egg in an amphibian, a bird, and a mammal. [8]
   b) the development of the embryo in an amphibian, a bird, and a mammal. [12]

2 Write notes on the following:
   a) allometric growth, [5]
   b) diapause, [5]
   c) senescence, [4]
   d) larval forms, [3]
   e) growth curves. [3]

3 a) Distinguish between growth and development. [4]
   b) Explain briefly how you could measure growth in seedlings. [6]
   c) Give an illustrated account of secondary thickening in the stem of a dicotyledon. [10]

4 a) Table 26.2 gives details of the growth of various tissues in man.

Table 26.2

Age in years	Size obtained as a percentage of total post-natal growth			
	Reproductive tissues	Lymphatic tissues	Brain and head	Overall growth
2	5	41	49	31
4	7	65	76	39
6	7	79	92	43
8	7	114	94	49
10	8	157	96	52
12	9	191	98	56
14	18	163	99	69
16	42	130	100	84
18	80	112	100	92
20	100	100	100	100

   (i) Express the results graphically. [6]
   (ii) Comment on the significance of the pattern of development shown by *each* line plotted on your graph. [8]
   b) It is possible to measure the overall growth of cress or oat seedlings in a variety of ways.
      (i) State *three* suitable methods.
      (ii) Briefly discuss the advantages and disadvantages of the three methods.
      (iii) Suggest which of the three would provide the most reliable data. [6]   (AEB)

5 Study the following percentage analysis (table 26.3) of the dry mass of seeds and seedlings taken over a period of early germination.

Table 26.3

	Ungerminated seeds %	5 day seedlings %	10 day seedlings %
Cellulose	15	15	20
Fat	33	19	15
Salts	5	5	5
Protein	24	22	23.5
Starch	0	7	3
Various organic	23	28	25.5
Loss in mass	0	4	8

a) (i) Suggest a possible explanation for the changing percentages of fat and starch and the relationship between these substances. [10]

(ii) Comment on *one* other change that you think significant. [10]

For germination to begin, factors other than adequate supplies of water and oxygen and normal temperatures may need to be present. Dormancy can be prolonged or broken by such factors as very low temperatures, or by the presence of chemicals in the fruit wall or perhaps others.

b) (i) Suggest the role played by very low temperatures in regulating the life cycles of plants under natural environmental conditions.

(ii) How might *either* low temperatures *or* chemicals in the fruit act to affect dormancy?

(iii) Light is also known to break dormancy in some seeds. Devise a simple experiment to determine the extent to which wavelength of incident light might be significant in breaking dormancy. [15]   (AEB)

27

The control of growth

Survey and question
Survey this chapter and make overview notes in pattern form.

Read
THE CONTROL OF PLANT GROWTH
EVIDENCE FOR THE INVOLVEMENT OF A HORMONE
AUXINS
RESPONSE TO GRAVITY
RESPONSE TO LIGHT

Recall

1   List the internal and external factors that influence growth in (a) animals, and (b) plants.

2   By means of annotated diagrams, show that:
    a) The tip of a growing shoot is responsible for causing growth to take place in the more posterior parts of the shoot.
    b) A diffusible chemical substance produced by the tip is responsible for causing growth in the more posterior parts of the shoot.
    c) The degree of curvature in a growing shoot is directly proportional to the concentration of growth substance present.
    d) Those concentrations of growth hormone that stimulate growth in shoots inhibit growth in roots.

3   a) Name the substance which promotes growth in stems.
    b) Where is this substance produced in growing plants?

4   a) Define tropism.
    b) How does a tropism differ from a taxis?

5   a) Explain tropism in roots and shoots in terms of auxin distribution.
    b) How may geotropism be demonstrated in the laboratory?

6   Describe two hypotheses to explain phototropism.

7   Describe experiments which support one of these hypotheses.

Read

OTHER RESPONSES
MODE OF ACTION OF AUXINS
OTHER AUXIN EFFECTS
GIBBERELLINS
KININS
ABSCISIN
ETHYLENE

Recall

1. **a)** Describe three responses to directional stimuli.
   b) Describe one response to non-directional stimuli in plants.
2. Copy and complete table 27.1.

Table 27.1 *A summary of the effects of plant hormones on growth and development.*

Effect on	Auxins	Gibberellins	Kinins	Abscisin	Ethylene
Stem: apical meristem					
Stem: secondary meristem					
Lateral buds					
Main root					
Adventitious roots					
Ovary wall					
Seeds					
Leaves					
Mode of action of hormone:					

3. Outline ways in which plant hormones can be used commercially by horticulturists.

Read

TEMPERATURE AND PLANT GROWTH
LIGHT AND PLANT GROWTH
PHOTOPERIODISM

Recall

1. List the effects of light on plant growth.

2. Describe the effect of the following sequences of light on the germination of lettuce seeds:
   a) far-red, red,
   b) far-red, red, far-red.

3. a) Name the pigment which absorbs these flashes of red and far-red light.
   b) Where is this pigment produced in plants?

4. Copy figure 27.1 and complete it by showing the interconversion of the forms of phytochrome during the day and night.

Figure 27.1 *Absorption of light by phytochrome*

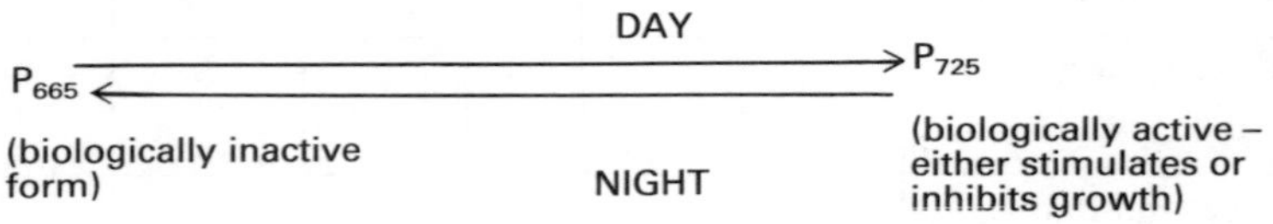

```
                        DAY
                ──────────────────────→   P₇₂₅
   P₆₆₅  ←──────────────────────
   (biologically inactive                (biologically active –
   form)                NIGHT             either stimulates or
                                          inhibits growth)
```

5. Copy and complete table 27.2

   Table 27.2

Effect on	Red light	Far-red light
Stem		
Leaves		
Lateral roots		
Flowering		

6. a) What is photoperiodism?
   b) Using named examples, describe how plants can be divided into groups on the basis of their response to periods of light and dark.

7 Copy and complete figure 27.2.

Figure 27.2 *Hypothetical scheme summarising the photoperiodic control of flowering*

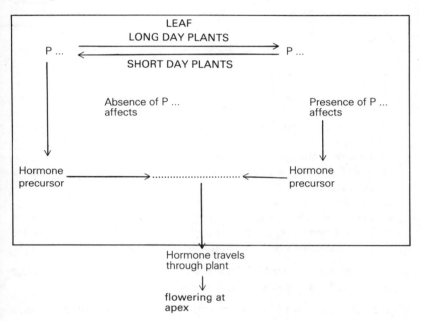

8 Describe the evidence which suggests that a hormone is involved in the photoperiodic control of flowering.
9 How do the following affect plant growth and development:
   a) an increase in temperature,
   b) exposure to cold?

Read
CONTROL OF GROWTH IN ANIMALS
ACTION OF A GROWTH HORMONE ILLUSTRATED BY INSECTS
METAMORPHOSIS

Recall
1 a) Name the site of production of growth hormone in mammals.
   b) How does growth hormone promote growth?

2 Copy and complete table 27.3.

Table 27.3 *A summary of the effects of abnormal production of growth hormone in humans*

Name of condition	Due to	Symptoms
Gigantism		
Acromegaly		
Dwarfism		

3 a) Name the hormone responsible for shedding of the cuticle in insects.
   b) Where is this hormone produced?
   c) This hormone is produced only at intervals. What triggers its production?
   d) Where is this produced?
   e) What evidence is there to suggest that a certain quantity of brain hormone is required before the thoracic gland can be stimulated to produce moulting hormone?

4 Describe evidence which indicates the mechanism by which the moulting hormone exerts its effects.

5 a) How does juvenile hormone influence the cuticle which develops after moulting?
   b) Name the gland responsible for juvenile hormone production, and explain its role in metamorphosis.

6 Summarise diagrammatically the role of the brain and the thoracic gland in the control of moulting and metamorphosis in an insect.

7 a) Name the hormone that controls metamorphosis in amphibians.
   b) Describe the mechanism by which this hormone is believed to exert its effects.

Read
DORMANCY AND SUSPENDED GROWTH
SURVIVAL DURING DORMANCY
THE MECHANISM OF DORMANCY

Recall
1. a) Explain the meaning of dormancy.
   b) Give some named examples of dormant stages in plants and animals.
   c) What are the advantages of dormancy?
2. List the factors that can cause dormancy in:
   a) seeds,
   b) buds,
   c) tubers and bulbs.
3. a) Define diapause.
   b) Explain how diapause is caused.
4. Describe (a) the similarities, and (b) the differences between hibernation and aestivation.

Review
1. a) Explain the meaning of metamorphosis. [4]
   b) Describe metamorphosis and its control in a named insect and an amphibian. [12]
   c) List the advantages of each of the above life cycles. [4]
2. Discuss how plants respond to light (exclude photosynthesis). [20]
3. a) What is meant by dormancy? [4]
   b) Discuss the biochemical aspects of dormancy in animals and plants. [16]
4. Survey the methods by which animals and plants are able to withstand unfavourable conditions. [20]
5. a) List the environmental factors that affect growth in organisms. [4]
   b) Discuss the role of growth regulating substances in plants and animals. [16]

28

Mendel and the laws of heredity

Survey and question
Survey this chapter and make overview notes in pattern form.

Read
MENDEL
MONOHYBRID INHERITANCE
CONCLUSIONS FROM THE MONOHYBRID CROSS

Recall
1. a) List the aims of the science of genetics.
   b) Name the person responsible for the first recorded studies of genetics.
2. a) Explain the meaning of monohybrid inheritance.
   b) How did Mendel obtain pure-breeding tall peas?
3. Explain the meaning of:
   a) first filial generation (F_1),
   b) second filial generation (F_2).
4. Copy and complete figure 28.1.

Figure 28.1 *Diagram summarising the results of some of Mendel's monohybrid crosses*

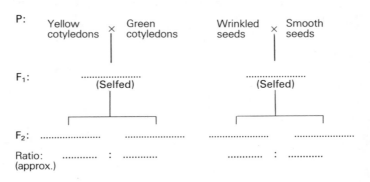

5 In the experiments summarised in question 4, explain why:
  a) the F_1 generations do not show features which are intermediate between those of each parent.
  b) only one parental feature shows up in each F_1 generation.
  c) both of the original (parental) features show up in the F_2 generation.

Read
GENES AND THEIR TRANSMISSION
IN TERMS OF PROBABILITY
THE ROLE OF CHANCE

Recall
1 Explain the meaning of: (a) gene, and (b) allele.

2 Construct genotypes of the following:
  a) homozygous pea plants with yellow cotyledons,
  b) heterozygous pea plants with yellow cotyledons,
  c) pea plants with green cotyledons.

3 Describe the phenotypes of pea plants with the following genotypes: (w = wrinkled seeds, W = smooth seeds). (a) Ww, (b) WW, (c) ww.
Which is the double recessive?

4 Construct diagrams to show the F_1 and F_2 genotypes produced by crosses between each of the following homozygous parents:
  a) wrinkled and smooth seeds,
  b) yellow and green cotyledons.

5 a) For each of the examples in question 4, construct a Punnet square to show the fusion of the F_1 gametes.
  b) Using the results from your Punnet squares state the probability of the occurrence of the following in the F_2 generations:
     (i) heterozygous smooth seeded offspring,
     (ii) homozygous smooth seeded offspring,
     (iii) wrinkled seeded offspring,
     (iv) phenotypes with green cotyledons,
     (v) phenotypes with yellow cotyledons.

Read
MENDEL'S FIRST LAW: THE LAW OF SEGREGATION
BREEDING TRUE
TEST CROSS
MONOHYBRID INHERITANCE IN MAN

Recall

1. **a)** State Mendel's first law (the law of segregation).
   b) Copy figure 28.2 and annotate it to explain Mendel's first law in terms of meiosis.

Figure 28.2

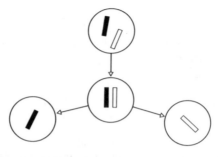

2. Which of the following represents a test cross (back cross)? Explain your answer.
   a) TT × Tt,
   b) TT × TT,
   c) TT × tt,
   d) Tt × tt.

3. Using genetic symbols, describe a test cross to identify true breeding long-winged *Drosophila*.

4. List some conditions in humans which are known to be associated with a single pair of genes.

5. What is the probability of
   a) the children of two achondroplastic dwarfs being of normal size.
   b) albinism developing in the children of a normal woman and a man heterozygous for albinism.

Read

DIHYBRID INHERITANCE
CONCLUSIONS FROM THE DIHYBRID CROSS
IN TERMS OF PROBABILITY
RELATIONSHIP BETWEEN GENOTYPE AND PHENOTYPE
TEST CROSS

Recall

1. Explain the meaning of dihybrid inheritance.

2 Construct genotypes for the following pea plants:
   a) homozygous plants with yellow cotyledons and smooth seeds,
   b) homozygous plants with green cotyledons and wrinkled seeds.
3 Make a diagram to show the F_1 and F_2 genotypes produced by crossing the two types of pea plants described in question 2.
4 What are the probabilities of the following occurring in the F_2 generation of the cross described in question 3:
   a) smooth-seeded plants with yellow cotyledons,
   b) wrinkled-seeded plants with green cotyledons,
   c) plants which are homozygous for yellow cotyledons and smooth seeds,
   d) smooth-seeded plants with green cotyledons,
   e) plants which are heterozygous for yellow cotyledons with smooth seeds?
5 By means of a Punnet square, show how you would carry out a test cross to identify true breeding smooth-seeded plants with yellow cotyledons.
6 State Mendel's second law.

Read
EXPLANATION OF MENDEL'S SECOND LAW
INDEPENDENT ASSORTMENT OF FRUIT FLY
MENDEL IN RETROSPECT

Recall
1 Copy figure 28.3 and annotate it to explain Mendel's second law in terms of meiosis.

Figure 28.3

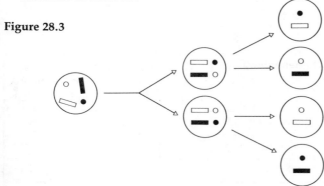

2 Make brief notes on the principles of scientific method shown by Mendel's research.

Review

1 a) Discuss the significance of Mendel's first law. [8]
   b) Using named examples, explain the meaning of the following:
      (i) alleles,
      (ii) dominance,
      (iii) genotype,
      (iv) test cross (back cross). [12]

2 Plants of *Mirabilis*, breeding true for both red flowers and broad leaves, were crossed with plants breeding true for both white flowers and narrow leaves. The resulting F_1 plants had pink coloured flowers and an intermediate leaf width. Suggest an explanation for this result.

   What would be the expected appearance of the plants produced (a) by crossing the F_1 plants amongst themselves, and (b) by crossing the F_1 plants with the red-flowered, broad-leaved parental type? Give reasons for your answer.

   Indicate how the results might have been different if the genes for flower colour and leaf width had been on the same chromosome. [20] (London)

3 a) A true breeding tall, purple-stemmed tomato plant is crossed with a dwarf green-stemmed tomato plant. What genotypes and phenotypes will be produced in: (i) the F_1, and (ii) the F_2 generations? [14]
   b) Table 28.1 gives results from a breeding experiment on tomato plants. What are the most probable genotypes of the parents in each case? [6]

Table 28.1

Parental phenotypes	Offspring phenotypes			
	Tall Purple	Tall Green	Short Purple	Short Green
Tall purple × Short green	301	287	293	305
Tall green × Short purple	521	0	0	0
Tall green × Short purple	87	91	77	83

29

Chromosomes and genes

Survey and question
Survey this chapter and make overview notes in pattern form.

Read
LINKAGE
LINKAGE GROUPS AND CHROMOSOMES
SEX DETERMINATION

Recall
1 Explain why the results of the *Drosophila* cross on page 462 in BAFA are unexpected.
2 Explain the meaning of (a) linked genes, and (b) linkage group.
3 What is the relationship between
   a) the number of linkage groups in an organism and the number of different types of chromosomes?
   b) the number of genes in a linkage group and the size of the chromosome concerned?
4 Define
   a) autosomes,
   b) heterosomes,
   c) homogametic,
   d) heterogametic.
5 Identify the type of sex chromosomes in the gametes of the following:
   a) male and female birds,
   b) male and female *Drosophila*,
   c) a male and female insect (other than *Drosophila*).

Read
SEX LINKAGE
SEX LINKAGE AND THE Y CHROMOSOME

Recall

1. State an example of sex linkage in (a) *Drosophila*, and (b) humans.
2. What are the possible results of a union between the following:
   a) a normal man and a woman carrying the red–green colour blindness gene,
   b) a white-eyed male *Drosophila* and a female red-eyed heterozygote,
   c) a red-eyed male *Drosophila* and a female red-eyed heterozygote?
3. Is it possible for a child of a normal man and a normal woman to be a haemophiliac? Explain your answer.
4. List the traits which are possibly carried on the Y chromosome in humans.

Read

CROSSING OVER
EXPLANATION OF CROSSING OVER

Recall

1. a) Explain the meaning of crossing-over.
   b) In a breeding experiment with *Drosophila*, the results in figure 29.1 were obtained.

Figure 29.1

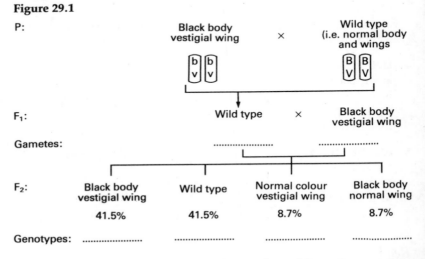

Copy this diagram and complete it by adding the possible gametes that could be produced by the F_1 flies and the genotypes of the F_2 flies.

2 By means of a diagram, show how meiosis in the F_1 can account for the genetic combinations found in the F_2 with reference to question 1.

Read
LOCATING GENES ON CHROMOSOMES

Recall
1 Explain the meaning of crossover frequency (crossover value COV).

2 What are the crossover frequencies of the genes in the *Drosophila* experiments (results of F_2 generations given) in figure 29.2?

Figure 29.2

a)
White-eye vestigial wing	Wild type	White-eye normal wing	Red-eye vestigial wing
32.5%	32.5%	17.5%	17.5%

b)
White-eye yellow body	Wild type	White-eye normal body	Red-eye yellow body
49.25%	49.25%	0.75%	0.75%

3 What do the results of these experiments in question 2 indicate about the relative position of the following genes on the chromosome:
  a) white eye and vestigial wings,
  b) white eye and yellow body?

4 Describe the cross you would have to carry out to determine the relative position of the genes for yellow body and vestigial wings.

5 Explain the meaning of (a) chromosome map, and (b) gene locus.

Read
CAN GENES BE SEEN?
MULTIPLE ALLELES
GENETICS AND THE LAW
DEGREES OF DOMINANCE
THE SICKLE-CELL GENE
GENE ACTION

Recall

1. a) How are giant chromosomes formed?
   b) Briefly summarise the information gained from studies of giant chromosomes.

2. Using a named example, explain the meaning of multiple allele.

3. Work out the genotypes that could result from the following unions:
   a) blood group A × blood group B
   b) blood group AB × blood group O.

4. Explain the results of the crosses in figure 29.3.

Figure 29.3

a) P:　　　　　　　　　red flowers × white flowers
　　F$_1$:　　　　　　　　　　　↓
　　　　　　　　　　　　　　pink flowers

b) P:　　　　　　　　blood group M × blood group N
　　F$_1$:　　　　　　　　　　　↓
　　　　　　　　　　　　blood group MN

5. A woman's first child suffered from sickle-cell anaemia.
   a) Indicate the baby's genotype and that of its parents.
   b) What is the probability that the second child will also suffer from the disease?

6. a) What evidence is there to show that the sickle-cell gene is not completely recessive to the gene for normal haemoglobin?
   b) Explain why people with the sickle-cell trait can live normal lives.

7. What is the difference between haemoglobin S and normal haemoglobin?

Read

LETHAL GENES
INTERACTION OF GENES

Recall

1. About one human being in a thousand shows a condition known as the Pelger anomaly, which is an abnormality of the white blood cells. The Pelger gene is dominant and inherited in the normal Mendelian manner. It has been shown, however, that people with the Pelger anomaly are always heterozygous; no homozygous individuals have ever been observed. Give a possible explanation for this.

2 Work out the results of crossing poultry with the following comb phenotypes:
   a) pea × single
   b) pea × rose
   c) walnut (homozygous) × walnut (heterozygous).
3 Explain the results of the above crosses.

Review

1 Using suitable named examples, explain the meaning of:
   a) crossing over, [5]
   b) incomplete dominance, [5]
   c) genetic interaction, [5]
   d) lethal genes. [5]
2 a) How is sex determined in (i) humans, and (ii) birds? [4]
   b) Using at least two named examples, explain the meaning of sex linkage (do not use the example given below). [6]
   c) The genes controlling eye colour in *Drosophila* are located on the X chromosome and are therefore sex linked. Explain clearly the results of crossing a white-eyed male with a red-eyed (heterozygous) female. [10]
3 a) A yellow cat gave birth to a litter of seven kittens; four yellow males and three tortoiseshell females. Assume there is a single father for the litter and give a full explanation of these results. [10]
   b) Predict the results of a cross between one of the female kittens and a black tomcat. [10]
4 a) What is meant by co-dominance? [6]
   b) The blood group of one of a pair of non-identical twins is A, whilst that of the other is blood group O. Give a full explanation of how this could occur. [14]
5 Draw up a comprehensive programme for a genetic investigation to determine whether two named morphological features that show discontinuous distribution in a given organism are linked or show independent assortment. Indicate the practical difficulties in carrying out such a programme. [20]   (AEB)

30

The nature of the gene

Survey and question
Survey this chapter and make overview notes in pattern form.

Read
EVIDENCE THAT NUCLEIC ACID IS INVOLVED
THE STRUCTURE OF NUCLEIC ACIDS

Recall
1. a) State two functions of genes.
   b) What properties of genetic material are necessary to carry out these functions?

2. a) What hypothesis did Avery and his colleagues test with their experiments on *Pneumococcus*?
   b) Describe their methods.
   c) Did the results support the hypothesis? Explain.

3. a) State the three structural components of a nucleotide.
   b) Using the symbols in figure 30.1, draw two adjacent, linked nucleotides.

Figure 30.1

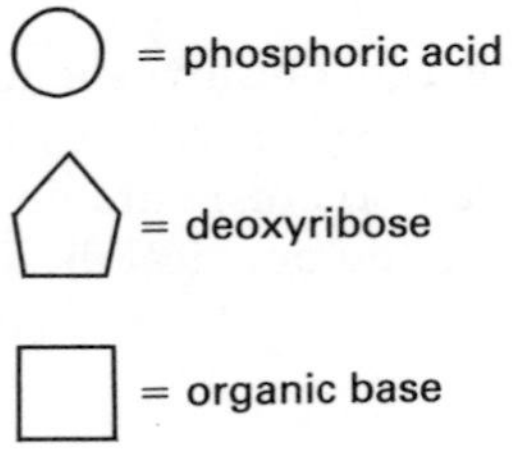

◯ = phosphoric acid

⬠ = deoxyribose

▢ = organic base

4. State how DNA differs chemically from RNA in terms of (a) the pentose sugars, and (b) the bases.

Read

THE WATSON–CRICK HYPOTHESIS
REPLICATION OF THE DNA MOLECULE

Recall

1 Describe the evidence which led Watson and Crick to suggest (a) that DNA consists of parallel strands held together by paired bases, and (b) that the DNA molecule is a 'double helix'.

2 a) Copy figure 30.2 and name the four different bases.

Figure 30.2 *The structure of DNA*

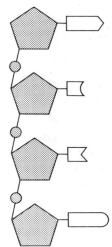

  b) Complete the diagram by drawing a complementary strand and label the bases.
  c) How are the bases linked together? Indicate this on your diagram.

3 a) Describe how Kornberg synthesized DNA.
  b) Explain how the results of his experiments indicated that DNA is capable of replication.

Read

CONSERVATIVE VERSUS SEMI-CONSERVATIVE REPLICATION
DNA REPLICATION AND MITOSIS

Recall

1 Copy figure 30.3 and annotate it to explain the replication of DNA.

Figure 30.3

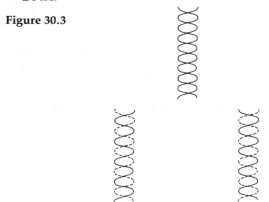

2 How do Meselson and Stahls' results support the semi-conservative hypothesis?

3 a) When does DNA replication occur in cells?
  b) Name the process by which replicated DNA is distributed equally between cells.
  c) Name the process by which the DNA component of cells is halved.

Read

THE ESSENTIAL ROLE OF DNA
DNA AND PROTEIN SYNTHESIS
DNA AS A CODE FOR PROTEINS
HOW DOES DNA COMMUNICATE WITH THE CYTOPLASM?
FORMATION OF MESSENGER RNA
THE ASSEMBLY OF A PROTEIN
THE ROLE OF THE RIBOSOMES

Recall

1 a) State the one gene–one enzyme hypothesis.
  b) How do the results of Beadle and Tatums' experiment support this hypothesis?

2 Name the following:
  a) the molecule responsible for communication between the DNA and the cytoplasm,

b) the molecule responsible for transferring amino acids from the cytoplasm to the site of protein synthesis,
c) the site of protein synthesis,
d) a triplet of bases specifying an amino acid.

3   a) Make a copy of figure 30.4.

Figure 30.4   *Diagram summarising the sequence of events in protein synthesis*

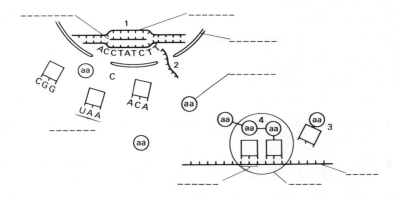

b) Label the structures indicated. Identify the base sequences on the DNA and RNA molecules (where possible).
c) Add notes to explain the events occurring in 1–4.

4   Describe the function of the following in protein synthesis:
a) ribosomes,
b) polysomes.

Read

EVIDENCE FOR DNA'S CONTROL OF PROTEIN SYNTHESIS
IS THE CODE OVERLAPPING OR NON-OVERLAPPING?
TRANSCRIPTION AND TRANSLATION OF THE GENETIC CODE
DNA'S CONTROL OVER THE CELL

Recall

1   Describe the evidence which indicates:
a) that DNA controls the synthesis of RNA,
b) the role of ribosomes in protein synthesis,
c) the role of RNA in the sequencing of amino acids in a polypeptide chain,

d) the role of transfer RNA in determining the sequence of amino acids in a polypeptide chain.

2 a) What is meant by genetic engineering?
   b) Describe some possible practical applications of genetic engineering.
   c) What are the dangers of genetic engineering?

3 In terms of protein synthesis, explain the meaning of:
   a) transcription,
   b) translation.

4 In terms of the genetic code, explain the meaning of:
   a) degenerate,
   b) nonsense triplets,
   c) overlapping code,
   d) non-overlapping code.

5 Describe the evidence which indicates that the code is non-overlapping.

6 State the DNA triplets necessary to code for a polypeptide chain containing the following sequences of amino acids:
   a) leucine, valine, serine, alanine,
   b) cysteine, histadine, phenylalanine, tryptophan.

7 a) What are bacteriophages?
   b) Describe the evidence from studies on bacteriophages that DNA is responsible for controlling protein synthesis.

Review

1 Survey the evidence which indicates that nucleic acid is the carrier of genetic information. [20]

2 Write notes on the following:
   a) nucleotides [5]
   b) nucleolus [5]
   c) ribosomes [5]
   d) anticodon [5]

3 a) Distinguish between bacteria and viruses. [8]
   b) Describe how studies on bacteria and viruses can help in understanding the mechanism of inheritance. [12]

4 a) What is meant by a mutation? [6]
   b) Describe, using named examples where possible, how mutations can occur. [14]

5 a) Describe those properties of DNA which make it suitable as heredity material. [8]
   b) By means of annotated diagrams, describe how this information is transcribed and translated. [12]

31

Genes and development

Survey and question
Survey this chapter and make overview notes in pattern form.

Read
THE IMPORTANCE OF THE NUCLEUS

Recall
1 a) Copy and complete figure 31.1.

Figure 31.1 *Hammerling's experiments with Acetabularia*

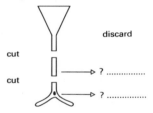

   b) How does this experiment support the following hypotheses?
   (i) The nucleus is required for cytoplasmic development.
   (ii) Some chemical substance passes from the nucleus into the cytoplasm, stimulating the formation of a new head.
   c) What further experiment would you carry out to test (ii)?

2 a) Copy and complete figure 31.2.

Figure 31.2 *Hammerling's experiments with Acetabularia*

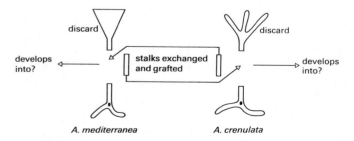

b) Explain the results of the experiment (figure 31.2) in terms of the control exerted by DNA and messenger RNA.

Read

THE PROBLEM OF MULTICELLULAR ORGANISMS
THE ROLE OF THE CYTOPLASM
THE FATE OF THE CELLS IN THE BLASTULA
TISSUE CULTURE AND GRAFTING
SPEMANN AND MANGOLD'S EXPERIMENT
ORGANIZERS
THE CHEMICAL NATURE OF ORGANIZERS
THE ROLE OF DNA IN DEVELOPMENT
CHROMOSOME PUFFS

Recall

1 **a)** Describe the evidence that indicates the role of the cytoplasm in differentiation.
   b) Although most of a cell's DNA is located in the nucleus, some can also be found in the cytoplasm. List those cytoplasmic structures which contain genetic material.

2 Describe the techniques that can be used to determine the following in the chordate embryo:
   a) the origin of the germ layers,
   b) whether the development of different areas of the embryo is determined from an early stage, or is dependent upon influences which arise as development proceeds.

3 Copy and complete table 31.1.

Table 31.1

Donor	Grafted onto	host	Develops into
a) Ectoderm	grafted onto	mesoderm	....
b) Mesoderm	grafted onto	ectoderm	....
c) Chorda	grafted onto	ectoderm	....
d) Ectoderm	grafted onto	chorda	....

4 Briefly explain the results in question 3 in terms of organizer, tissues and induction.

5 Explain the following observations.
   a) If the developing optic vesicle is carefully removed from a vertebrate embryo, the lens will not develop.

b) If a developing optic vesicle from a vertebrate embryo is transplanted into the embryonic epidermis, an ectopic eye develops.
c) The nucleus taken from a differentiated cell in a vertebrate embryo can, when transplanted into an egg, bring about the development of a normal embryo. In certain cases, development is abnormal.

Read

GENE SWITCHING
THE JACOB–MONOD HYPOTHESIS OF GENE ACTION
THE POSSIBILITY OF CYTOPLASMIC CONTROL
THE ROLE OF THE ENVIRONMENT
FORMATION OF RIBOSOMES: THE ROLE OF THE NUCLEOLUS
SENESCENCE

Recall

1 With reference to human haemoglobin, explain the meaning of gene switching.

2 a) Copy and complete figure 31.3.

Figure 31.3 *The Jacob–Monod Hypothesis of gene action*

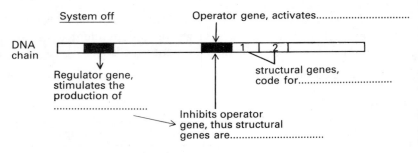

b) Construct a similar diagram to show what happens when enzyme is required (i.e. system on).
c) What experimental evidence led Jacob and Monod to propose this hypothesis?

3 Describe evidence which suggests that protein synthesis
a) can be controlled by the cytoplasm,
b) is influenced by the external environment.

4 a) What is the function of the nucleolus?
b) Briefly describe how this function is believed to be carried out.

5 a) What is senescence?
   b) List some of the symptoms of senescence in humans.

Review

1 Discuss the origin, structure and function of the extra-embryonic membranes in a bird and a mammal. [20]

2 a) What is meant by development? [5]
   b) Describe the experiments that indicate how development is brought about in a named organism. [15]

3 a) With reference to a named organism, write brief notes on the following:
      (i) fertilization,
      (ii) cleavage. [10]
   b) (i) What is meant by an organizer?
      (ii) Discuss the chemical nature of organizers. [10]

4 a) What is meant by senescence? [4]
   b) Give a full account of some of the possible reasons for senescence. [16]

5 a) What is meant by genetic engineering? [4]
   b) Discuss some of the possible benefits and dangers of genetic engineering. [16]

32

The organism and its environment

This is the first chapter in part VI: 'Ecology and evolution'. The overview notes below show how the topics covered by the chapters in this part of the book are related.

Overview notes for part VI: Ecology and evolution
(Chapter numbers given in brackets)

Figure 32.1

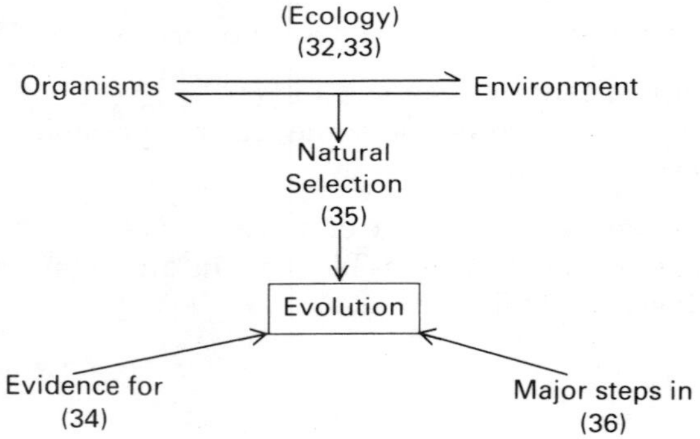

```
                    (Ecology)
                    (32,33)
    Organisms  ⇌  Environment
                       ↓
                   Natural
                   Selection
                    (35)
                       ↓
                   Evolution
              ↙                ↘
     Evidence for          Major steps in
         (34)                   (36)
```

Survey and question
Survey this chapter and make overview notes in pattern form.

Read
FROM BIOSPHERE TO ECOLOGICAL NICHE
THE PHYSICAL ENVIRONMENT
THE BIOTIC ENVIRONMENT
THE CONCEPT OF THE ECOSYSTEM

Recall
1   Copy and complete figure 32.2.

Figure 32.2 *Ecological levels of organisation*

2   Write brief explanations of the terms: (a) environment, and (b) ecological niche.

3   Summarise the main environmental factors in tabular form as shown in figure 32.3.

Figure 32.3   ENVIRONMENT

```
        PHYSICAL                    LIVING
       (..............)           (..............)
    1                            1
    2    etc.                    2    etc.
```

4   Copy and complete the following definitions:
   a) A population is a number of individuals of the same .... occupying the ....
   b) A .... is a number of populations of different .... occupying the ....
   c) The community, along with its non-living environment, is called the ....

Read

POPULATION GROWTH
ENVIRONMENTAL RESISTANCE
MAINTENANCE OF POPULATIONS
SUDDEN CHANGES IN POPULATIONS
CONTROL OF THE HUMAN POPULATION

Recall

1   Draw a graph to show the growth in population of organisms in a previously unoccupied environment. (Assume no predators

and an ample food supply.) By means of vertical lines indicate the four main phases in the development of the population. Use your answers to the questions below to annotate the graph.

a) Initially population growth is slow. Why?
Copy and complete the following questions.

b) When the population is increasing at its maximum rate:
   (i) The population increases ....
   (ii) The .... of the population is at its maximum.
   (iii) There is no ....

c) The growth rate decreases because ....
   .... increases, causing either a decrease in .... or an increase in ...., or both.

d) When the graph reaches a plateau:
   (i) .... is at its maximum.
   (ii) The full .... of the environment has been reached.
   (iii) .... equals ....

2 List the main components of environmental resistance.

3 Draw a graph to show the population changes you would expect to find if a small number of rabbits and foxes were introduced to a grassy island previously unoccupied by either species.

4 a) Draw a graph to show the changes in the deer population of the Kaibab plateau between 1900 and 1935. Assume that the final population norm was 5000. Annotate the graph to indicate the major components of environmental resistance (i.e. limiting factors) prior to 1907 and after 1925.

  b) Draw a second graph to show the changes in environmental resistance (arbitrary units) that occurred over the same period.

  c) After the population crash of 1925 the population reached a new norm not much greater than the original 1907 norm, and did not increase beyond this. Suggest an explanation for this.

5 List six different methods of birth control.

6 Using the data given in figure 32.10 in BAFA, calculate the percentage increase in the human population by the year 2020 as compared to 1969, according to the three projections **A**, **B** and **C**.

Read
SUCCESSION
THE CYCLING OF MATTER AND THE FLOW OF ENERGY

Recall
1 Copy and complete figure 32.4.

Figure 32.4

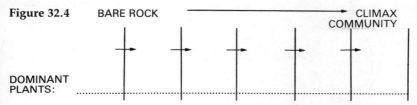

2 What is the major abiotic factor which influences the changes in vegetation described in question 1?
3 List the three nutritional groups in ecosystems, giving an example of each.
4 Explain briefly what would happen if all the decomposers were removed from an ecosystem.
5 Explain briefly why it would be incorrect to speak of an energy cycle in the biosphere.

Read
THE CARBON, OXYGEN AND NITROGEN CYCLES

Recall
1 Copy and complete figure 32.5.

Figure 32.5 *The carbon cycle*

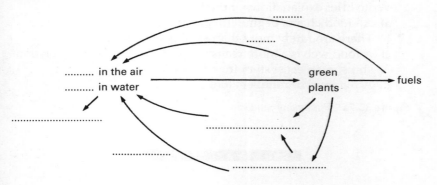

2 Construct a similar diagram to illustrate the oxygen cycle.
3 Copy and complete figure 32.6.

Figure 32.6 *The nitrogen cycle*

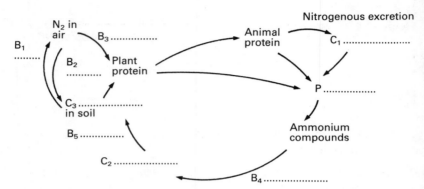

(B_1–B_5 represent types of bacteria, C_1–C_3 are chemical compounds and P is a process.)

4 Refer back to BAFA, page 159, and also page 192, and give *named* examples of each of the different types of bacteria B_3, B_4 and B_5 in the diagram above.

Read
FOOD CHAINS AND FOOD WEBS
PYRAMIDS OF BIOMASS AND NUMBERS

Recall
1 Define biomass.
2 Write brief explanations of the following.
   a) All food chains begin with an autotroph.
   b) There are rarely more than six links in a food chain.
   c) A food web is a more realistic description of feeding relationships in an ecosystem than a food chain.
   d) A pyramid of numbers looks like this:

Figure 32.7a *Pyramid of numbers*

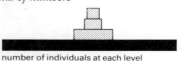

number of individuals at each level

but a pyramid of biomass for the same ecosystem looks like this:

Figure 32.7b *Pyramid of biomass*

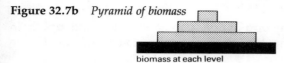

biomass at each level

3 Consider an ecosystem based on an oak tree. Draw (a) a pyramid of numbers, and (b) a pyramid of biomass for this ecosystem.

4 Is it possible for an organism to occupy more than one trophic level (energy level) at the same time? Explain your answer.

Read

AUTECOLOGY
RHYTHMS AND THE ENVIRONMENT
EXPLOITATION AND CONSERVATION

Recall

1 Write contrasting definitions of
   a) autecology,
   b) synecology (you may need to refer back to BAFA, page 524).
2 Explain, giving three examples for each: (a) diurnal rhythms, and (b) annual rhythms.
3 What is the difference between diurnal and circadian rhythms?
4 Write brief explanations, with examples of:
   a) carrying capacity,
   b) over exploitation,
   c) conservation.

Read

THE EFFECT OF MAN ON THE ECOSYSTEM
POLLUTION

Recall

1 What is the effect on an ecosystem of exterminating one of the species in it?
2 a) Identify two disadvantageous effects on an ecosystem of the use of pesticides.
   b) When a farmer uses chemical pesticides, the aim is to completely eradicate the pest. Is this true for biological control methods?
   Explain your answer.

3 Write brief notes to explain how the use of fertilizers can have harmful effects on aquatic ecosystems.
4 Explain briefly why bird of prey populations declined rapidly in the 1960s while smaller birds were not seriously affected by the use of pesticides.
5 Define pollution.
6 Construct a table to show six different types of pollutant and their main effects.

Review

1 a) List the main sources of environmental pollution. [4]
   b) Describe the effects of pollution on living organisms. [12]
   c) How may pollution be controlled? [4]
2 Study figures 32.8 and 32.9 which show the population changes of three organisms.

Figure 32.8

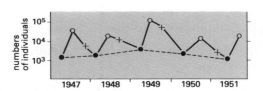

Annual population cycle of a grasshopper on a site in southern England. ● = adults, ○ = eggs, + = nymphs

Figure 32.9

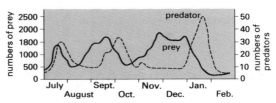

Population changes of the predatory mite *Typhlodromus* and its prey the plant mite *Eotetranychus*.

a) Outline the possible causes for fluctuation in the population.
   (i) from year to year in figure 32.8,
   (ii) throughout the year in figure 32.9.
b) Suggest the probable long term effects on the total numbers of adults in each population.
c) Explain how the biological control evident in figure 32.9 differs from the use of pesticides in its long term effect.

d) Plan a research programme which would determine the effect of removing the predatory mite from the location of the plant mite. [20]
(AEB)

3 Some authorities believe that, when an area of bare ground occurs in a community such as pasture, the organisms which recolonise the ground come in from outside by invasion. Others claim that the propagules (seeds, spores etc.) present in the soil itself are mainly responsible for the recolonisation.

Carefully describe how you would carry out a series of experiments to determine the validity of these two theories. [20]
(AEB)

4 The data in figure 32.10 show the community pyramids for an experimental pond. Productivity was estimated from rate of phosphorus uptake. Width of steps for numbers of organisms are on a logarithmic scale.

Figure 32.10

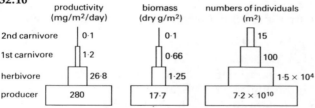

a) Compare the relative merits of the three different methods of representing the community pyramids.
b) Explain why the number of steps in the pyramids is usually restricted to four or five.
c) Suggest how the three pyramids would differ if they represented an oak wood, or other named deciduous wood. [20]
(AEB)

5 a) Using named examples, distinguish between each of the following:
    (i) food chain and food web,
    (ii) ecological niche and habitat,
    (iii) edaphic factor and biotic factor. [9]
b) Give an account of an investigation you have carried out on a named ecosystem. [11]

33

Associations between organisms

Survey and question
Survey this chapter and make overview notes in pattern form.

Read
PARASITISM
ADJUSTMENT BETWEEN PARASITE AND HOST
PARASITIC ADAPTATIONS
PARASITES AND MAN

Recall
1. What are the essential features of a parasitic relationship?
2. Write notes on the following:
   a) a plant parasitic on a plant,
   b) an endoparasite found in the bloodstream,
   c) an intracellular parasite (animal),
   d) an endoparasite found in the lymphatic system,
   e) an endoparasite living in the gut (unattached to host),
   f) an endoparasite living in the gut (attached to host's gut wall),
   g) the liver fluke.

   Organise your notes under the following headings:
   (i) benefits gained by parasite,
   (ii) harm suffered by host,
   (iii) adaptations of parasite to mode of life,
   (iv) methods of control.

Read
COMMENSALISM
SYMBIOSIS
SOME QUESTIONS CONCERNING THESE ASSOCIATIONS

Recall
1 What are the essential features of (a) commensalism, and (b) symbiosis?

2 Make notes on the following:
   a) a commensal relationship,
   b) a symbiotic relationship between two plants,
   c) a symbiotic relationship between two animals,
   d) a symbiotic relationship between a plant and an animal.

Read
SOCIAL ANIMALS
INTERACTION BETWEEN INDIVIDUALS

Recall
1 Study figure 33.16 in BAFA carefully, and reproduce it in your notes preferably from memory.

2 List three examples of ants making use of other organisms.

3 By means of annotated diagrams, describe and explain the possible functions (according to von Frisch) of the 'dances' performed by honeybees.

4 What is the most obvious difference between social behaviour in mammals and that of insects?

Review
1 In the biological control of insect pests, when parasites or predators of the pest species are introduced to areas where the pest is a problem, it is often found that the most successful control organisms have the following characteristics:
   a) their life cycle is of the same duration as that of the pest,
   b) they originate from areas of approximately the same latitude as the pest,
   c) they attack only the pest species,
   d) they attack late, rather than early, life cycle stages of the parasite.

   Explain why you think that these features are desirable in a biological control agent.

   Even though a parasite or predator with these characteristics may prevent excessive outbreaks of the pest it will be unable to prevent periodic rises in pest density. Why is this and what features would you look for in seeking a second control organism to supplement the first? [20] (AEB)

2 Using named examples, distinguish between:
   a) parasitism and predation, [6]
   b) commensalism and symbiosis, [6]
   c) saprophytism and scavenging. [8]

3 a) With reference to the *Hymenoptera*, explain the meaning of a social life in animals. [8]
   b) Compare and contrast the type of society in social insects with that of a named primate. [12]

4 a) List three parasitic diseases in humans. [3]
   b) Describe the life cycle of one of these parasites and the symptoms of the disease it causes. [12]
   c) How may this parasite be controlled? [5]

34
Evolution in evidence

Survey and question
Survey this chapter and make overview notes in pattern form.

Read
DARWIN AND THE THEORY OF EVOLUTION
WHAT DID DARWIN DO?

Recall
1 What is the significance of the following dates in Darwin's life? 1832, 1858, 1859.
2 Why did Darwin's theory bring him into conflict with the church?
3 What were the two major contributions Darwin made to biology?
4 List six areas of biology which provide evidence that evolution has occurred.

Read
DISTRIBUTION STUDIES
CONTINENTAL DISTRIBUTION
EXPLANATION OF CONTINENTAL DISTRIBUTION
EVIDENCE FOR MIGRATION AND ISOLATION
THE AUSTRALIAN FAUNA
CONTINENTAL DRIFT
OCEANIC ISLANDS
THE GALAPAGOS ISLANDS
DARWIN'S FINCHES
ISOLATING BARRIERS

Recall

1 Make a table with three columns headed Africa, South America, Australia, and list the characteristic groups of mammals found in each.

2 In what way does the mammalian fauna of the northern hemisphere differ from that of the southern hemisphere?

3 Write a concise summary of the hypothesis to explain the distribution of the mammalian fauna presented on pages 562–3 in BAFA. This hypothesis is based on there being four land bridges connecting the continents in the past. List these, and explain how the continents of the southern hemisphere are believed to have subsequently become isolated.

4 How does the distribution of the Camelidae lend support to this hypothesis?

5 In note form, explain how the Australian mammalian fauna is evidence for a process of adaptive radiation.

6 a) Write a concise summary of the theory of continental drift.
   b) Who originally put forward this theory, and when?
   c) What is the modern name for continental drift?
   d) How does the distribution of *Mesosaurus* support the theory?
   e) What is the name given to the original large land mass?
   f) Why did continental drift not affect the distribution of the placental mammals?

7 In note form, explain how the existence of the different species of Darwin's finches in the Galapagos islands may be explained in terms of adaptive radiation.

Read

COMPARATIVE ANATOMY
THE PENTADACTYL LIMB
DIVERGENT EVOLUTION
RECONSTRUCTING AN EVOLUTIONARY PATHWAY: THE VERTEBRATE
  HEART AND ARTERIAL ARCHES
CONVERGENT EVOLUTION

Recall

1 What is meant by homologous organs?

2 Make a diagram of a generalised pentadactyl limb.

3 Copy and complete table 34.1 to show how the basic pentadactyl limb pattern may be modified.

Table 34.1

Animal	Modification	Function
Monkey	Digits elongated	Grasping
Pig	etc.	etc.
Horse		
Mole		
Anteater		
Whale		
Bat		

4 Give two examples of vestigial structures in animals.

5 What is the name given to the process which results in adaptive radiation?

6 Figure 34.1 is a simplified version of figure 34.12 in BAFA, using single lines to represent the blood vessels. Make a copy of it, label and annotate it. Make similar diagrams to show the heart and arterial arches in an adult amphibian, a crocodile, a bird and a mammal.

Figure 34.1

7 a) What conclusions about the possible evolutionary relationships of the vertebrates can be drawn from a comparative study of the heart and arterial arches?

  b) How should these conclusions be qualified in the light of fossil evidence?

8 Copy and complete table 34.2 (p.222) to compare and contrast analogous structures with homologous structures. (Refer back to your answers to questions 1 and 5 if necessary.)

Table 34.2

	Analogous	Homologous
Definition		
Example		
Results from		

Read

TAXONOMY – THE CLASSIFICATION OF ORGANISMS
THE BASIS OF A NATURAL CLASSIFICATION
CLASSIFICATION OF THE CHORDATES

Recall

1 There are many ways in which plants may be classified. Here are some possible major groupings:
   a) annuals/perennials,
   b) fruits/vegetables,
   c) flowering/non-flowering,
   d) trees/herbs.
   Which of these classifications is the most 'natural' and which is the least 'natural'? Explain your answer.

2 List five features characteristic of all chordates. Which of these is used as the basis for the division of the phylum into subphyla?

3 Assuming that the classification of the chordates reflects their evolutionary relationships, arrange the following pairs in *decreasing* order of similarity:

$$\text{Most closely related} \\ \downarrow \\ \text{Least closely related}$$

   a) a frog and a sparrow,
   b) amphioxus and a shark,
   c) an eagle and a lion,
   d) a whale and a seasquirt.

Read

EMBRYOLOGY
THE TADPOLE LARVA OF SEA SQUIRTS
THE CONNECTION BETWEEN ANNELIDS AND MOLLUSCS
EVOLUTION AND EMBRYOLOGY: A WARNING

Recall

1 In note form, describe one example of embryological studies providing evidence for evolutionary relationships.

2 Explain in your own words Haeckels Recapitulation Theory: 'ontogeny recapitulates phylogeny'. Give two objections to the theory.

Read

CELL BIOLOGY
DISTRIBUTION OF PHOSPHAGENS
DISTRIBUTION OF BLOOD PIGMENTS
SEROLOGICAL TESTS

Recall

1 Make a completed version of table 34.3. Indicate presence or absence of a phosphagen by + or −.

Table 34.3

	Annelids	Molluscs	Arthropods	Echinoderms	Chordates
Phosphagens Found in .... tissue. Function .... 2 Types					
1...............					
2...............					

2 List the steps in the process of carrying out a serological test, the observations to be made, and the significance of the results.

Read

PALAEONTOLOGY
HOW FOSSILS ARE FORMED
RECONSTRUCTING EVOLUTIONARY PATHWAYS FROM THE FOSSIL RECORD
DATING FOSSIL REMAINS
THE EVOLUTION OF HORSES
EVOLUTION OF VERTEBRATE EAR OSSICLES

Recall

1 List three types of sedimentary rock, explaining their origin.

2 Describe in note form (a) the three main ways in which fossils may be formed, and (b) three other ways in which plant and animal remains may be preserved.

3 Complete figure 34.2 to summarise the information given in table 34.1 in BAFA. Indicate the three eras in the left-hand column, and the appearance of the groups of organisms listed along the bottom by means of a vertical line, as shown for the ferns. Use a dotted line for aquatic organisms and a solid line for terrestrial organisms.

Figure 34.2

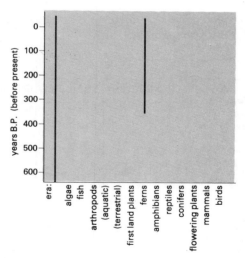

4 Which dating method would you use to estimate the age of a fossil dinosaur bone? Give reasons for your answer.

5 In note form, outline the differences in feet, teeth and size of the various forerunners of the modern horse. (There is no need to draw detailed diagrams of the teeth or skeleton.)

Review

1 Discuss how studies of each of the following have contributed to an understanding of evolution:
   a) Australasian mammals, [10]
   b) Galapagos finches. [10]

2 Account for the following:
   a) Most of the living fossils have been found in the sea. [5]
   b) Most mutations are not beneficial. [5]
   c) Continued inbreeding can be harmful. [5]
   d) The hind limbs of rabbits and birds have a similar basic structure. [5]

3 a) List the main types of studies that have provided evidence for evolution. [6]
   b) Describe how two of these studies have contributed to our understanding of evolution. [14]

4 a) Explain the meaning of taxonomy. [4]
   b) Using named examples, explain the basis of a natural classification. [16]

5 a) What is a fossil and how are fossils formed? [8]
   b) Using named examples, describe how a study of fossils can contribute to evolutionary theory. [12]

35
The mechanism of evolution

Survey and question
Survey this chapter and make overview notes in pattern form.

Read
SUMMARY OF DARWIN'S THEORY
LAMARCK'S THEORY
THE DARWINIAN AND LAMARCKIAN THEORIES COMPARED

Recall
1 What type of variation is exemplified by:
   a) Mendel's pea plants (see page 449),
   b) height in human beings?
2 Explain fully, in note form, the meanings of the following:
   a) survival of the fittest,
   b) natural selection,
   c) transmission of acquired characteristics.

Read
VARIATION
THE CAUSES OF GENETIC VARIATION
RESHUFFLING OF GENES
THE ROLE OF RESHUFFLING OF GENES IN EVOLUTION

Recall
1 How would a frequency distribution histogram for height in a population of Mendel's peas (see page 449) differ from that for height in human beings?
2 A pea plant with the genotype **TT** grown in a wind tunnel is short and stunted. If it is 'selfed' and the seeds collected and grown under normal conditions, what proportion of the F_1 generation will be short and stunted? Explain your answer.

3 Explain what is meant by:
   a) polygenic character,
   b) modifier gene.

4 Figure 35.1 shows two cells, containing one pair of homologous chromosomes, in the process of undergoing meiotic division to form gametes. In (b) a crossover has occurred between the **A** and **B** genes. Draw diagrams to show the different types of gametes that will result in each case. Which parent (a or b) will produce the most varied offspring?

Figure 35.1

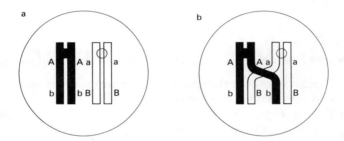

5 Explain briefly why the variation resulting from the reshuffling of genes in meiosis plays only a limited role in evolution.

Read

WHAT ARE MUTATIONS?
THE SPREAD OF MUTATIONS THROUGH A POPULATION
THE TYPES OF MUTATION
CHROMOSOME MUTATIONS

Recall

1 Briefly explain, and give examples of:
   a) mutation frequency,
   b) mutagenic agents.

2 List three characteristic features of mutations.

3 Write brief notes to explain the following:
   a) the slow spread of recessive mutations through a population,
   b) the complete disappearance of a mutation from a population,
   c) semi-dominant mutant genes.

4 *Without referring back to figure 35.4 in BAFA*, make annotated diagrams to illustrate the following types of chromosome mutation:
   a) deletion,
   b) inversion,
   c) translocation,
   d) duplication.

5 Write notes to explain, with examples, the meaning and significance of non-disjunction.

Read
POLYPLOIDY

Recall
1 Describe two ways in which the polyploid condition may arise.

2 a) Explain why offspring produced as a result of a cross between two different species are usually sterile.
   b) Explain how the development of polyploidy may result in a sterile hybrid becoming fertile.

3 Construct a diagram to show the cross which resulted in the appearance of *Spartina townsendii*. Indicate on your diagram the diploid and haploid numbers of the parent species, and the expected and actual chromosome number of the hybrid.

4 Why is polyploidy in plants of great importance to man? How can polyploidy be artificially induced?

Read
GENE MUTATIONS
DIFFERENT KINDS OF GENE MUTATIONS
THE IMPORTANCE OF MUTATION IN EVOLUTION

Recall
1 a) Explain the essential difference between chromosome and gene mutations.
   b) List, with brief explanations, four types of gene mutation.

2 Which type of mutation is likely to have been of most importance in evolution? Explain your answer.

Read

THE STRUGGLE FOR EXISTENCE
POPULATIONS AND EVOLUTION
NATURAL SELECTION
THE ACTION OF NATURAL SELECTION ON GENES
NATURAL SELECTION AS AN AGENT OF CONSTANCY AS WELL AS CHANGE

Recall

1 Define differential mortality.
2 Explain, briefly, the two ways in which genes may be lost from a population.
3 Write notes on sickle-cell anaemia and sickle-cell trait. (Refer back to pages 477 and 599 as well as 602.)
4 Under what conditions do (a) stabilizing, and (b) progressive selection operate?

Read

NATURAL SELECTION IN ACTION

Recall

1 It has been observed that the melanic form of the moth *Biston betularia* is far more abundant in industrial regions than the normal form, which is far more abundant in rural areas than the melanic form. Formulate and write out an hypothesis to explain these observations, and describe in note form an experiment designed to test the hypothesis.
2 In what way did the feeding habits of thrushes help Cain, Curry and Sheppard in their investigations into the distribution of *Cepaea*?
3 a) What is meant by area effects?
   b) List three possible explanations of area effects.

Read

POLYMORPHISM
NATURAL SELECTION AND POPULATION GENETICS
THE EFFECT OF ISOLATION ON GENE FREQUENCY

Recall

1 Describe examples of four different types of polymorphism.

2 Briefly explain what is meant by the following:
   a) deme,
   b) gene pool,
   c) gene frequency,
   d) Hardy–Weinberg principle.

3 List and briefly explain four ways in which the genetic equilibrium of a deme may be upset.

Read
APPLICATION OF THE HARDY–WEINBERG PRINCIPLE

Recall
1 a) List the assumptions made in deriving the Hardy–Weinberg formula.
   b) State the Hardy–Weinberg formula, and indicate the meaning of the symbols.
   c) Under what circumstances would you expect the genotype frequencies in a population to change?

Read
THE ORIGIN OF SPECIES
THE IMPORTANCE OF ISOLATION
ISOLATING MECHANISMS
THE EMERGENCE OF NEW SPECIES

Recall
1 Define species.
2 List the three types of isolating mechanisms, with examples.
3 Explain the meanings of: (a) allopatric, and (b) sympatric.
4 Which of the following pairs of gulls are capable of interbreeding successfully?
   a) lesser black-backed × Siberian Vega.
   b) lesser black-backed × American herring.
   c) American herring × British herring.
   d) British herring × Scandinavian black-backed.
   e) British herring × lesser black-backed.

Read
EVOLUTION IN RETROSPECT
ARTIFICIAL SELECTION

Recall

1 The rate of evolution of new species may have fluctuated dramatically in the past. Describe two factors which may have produced an increased rate of evolution.

2 Explain briefly why inbreeding can lead to unhealthy individuals whereas outbreeding can result in hybrid vigour.

3 List some of the qualities a plant breeder hopes to develop in a crop plant.

Review

1 a) What do you understand by speciation? [5]
  b) Discuss some of the factors and conditions which could bring about speciation. [15]

2 a) Describe Darwin's theory of evolution. [10]
  b) How does Darwin's theory differ from Lamarck's? [4]
  c) Discuss the current controversy surrounding Darwin's theory. [6]

3 The peppered moth, *Biston betularia*, is common in England. Typically it has speckled white wings and rests on the trunks and branches of trees. It is preyed on by birds which pick it from the trees on which it sits. In 1848 a dark melanic form was reported from the industrial area of Manchester where the bark of the trees was blackened by soot, and since that time the melanic form has increased in numbers, particularly in the industrial regions of the country.

In the 1950s Dr H.B.D. Kettlewell released a number of inconspicuously marked melanic and white male moths into woodlands in an industrial part of Birmingham and also in a rural part of Dorset. Later he attempted to recapture the moths using trapped virgin females and mercury vapour lamps. The results of his experiments are set out in table 35.1.

   a) From the information given above how would you explain the results of Dr Kettlewell's experiment? [6]
   b) State *three* factors which could have affected the accuracy of the results and what steps you could take to avoid them. [5]
   c) State precisely Darwin's principles of evolution by natural selection. Where appropriate use the data from the experiments with the moths to illustrate the principles you have stated. [5]

Table 35.1

		White	Melanic
Dorset 1955	released	496	473
	recaptured	62	30
	percentage recaptured	12.5	6.3
Birmingham 1953	released	137	447
	recaptured	18	123
	percentage recaptured	13.1	27.5

d) What is the most likely explanation for the appearance and subsequent survival in Manchester of the first melanic forms in 1848? [2]

e) Recent surveys show a steady increase in the typical white form in the Manchester area in the last decades. Suggest a reason for this change. [2]

f) Briefly describe *one* example of any other experimental work on evolution with which you are familiar. [5] (AEB)

4 The following is an extract from an article on the evolution of species:

'It should be stressed that evolution does not proceed at a constant rate. Its rate depends upon the relative constancy or inconstancy of the environment; the rate of mutation; the size of the breeding population; the intensity of competition; the intensity of selection pressure and the variety of selection pressures. The most slowly evolving organisms are those which inhabit relatively stable environments such as the open sea. The more intense the competition and selection pressure and the higher the rate of mutation, the more rapidly will evolutionary change proceed. Finally, the smaller the breeding population the more rapidly can innovations be tested and either eliminated or fixed in the gene pool of the population.' (N.M. Jessop, Biosphere; A Study of Life.)

a) In the context of the above passage explain briefly by means of theoretical or actual examples the meaning of the following: *gene pool; environment; mutation; competition; selection pressure; population.*

b) In what ways can the sea be said to be a more stable environment than most terrestrial environments? (Oxford)

36

Some major steps in evolution

Survey and question
Survey this chapter and make overview notes in pattern form.

Read
REFLECTIONS ON THE ORIGIN OF LIFE
SYNTHESIS OF ORGANIC MOLECULES

Recall
N.B. *The questions that follow relate to a hypothetical sequence of events.*

1. a) Write a concise statement of the spontaneous generation theory.
   b) In note form, summarise Pasteur's experiment of 1862 under the headings: method, results and conclusions.

2. a) What is the estimated age of the earth?
   b) When were the first indications of life?
   c) When was the probable origin of life?

3. a) List four constituents of the earth's primitive atmosphere.
   b) Explain why gaseous oxygen could not have existed in the primitive atmosphere.
   c) What was the probable source of the water vapour in the early atmosphere?

4. a) List three sources of the energy necessary for the synthesis of organic molecules on the primitive earth.
   b) From memory, make a simple diagram of Miller's apparatus.
   c) Copy and complete table 36.1, overleaf.

Table 36.1	Miller's Experiment	Calvin's Experiment
Energy source		
Products		

Read
THE FIRST ORGANISMS

Recall
1 What are the two key properties that distinguish living from non-living organic matter?

2 Were the first living organisms (a) autotrophic or heterotrophic, (b) aerobic or anaerobic? Give reasons for your answer.

Read
DEVELOPMENT OF AUTOTROPHS
AEROBIC RESPIRATION AND THE DEVELOPMENT OF SECONDARY HETEROTROPHS
THE DIVERGENCE OF ANIMALS AND PLANTS

Recall
1 a) What conditions made the evolution of autotrophic organisms increasingly likely?
   b) Write brief definitions of the two forms of autotrophic nutrition.

2 List three consequences of the evolution of photosynthetic autotrophs.

3 What was the main difference between the first heterotrophic organisms which began to decline after the appearance of autotrophs, and the heterotrophs which emerged later?

4 Name, and briefly describe, an organism capable of both autotrophic and heterotrophic nutrition.

Read
THE PROBLEM OF BACTERIA AND VIRUSES

Recall

1 Construct a table, as shown in table 36.2, to summarise the main differences between akaryotes, prokaryotes and eukaryotes.

Table 36.2

Akaryotes	Prokaryotes	Eukaryotes
1		
2 etc.		

Read

ORIGIN OF MULTICELLULAR ORGANISMS
LARVAL FORMS AND EVOLUTION

Recall

1 Summarise two alternative hypotheses to explain the origin of multicellular organisms, and describe one piece of indirect evidence in support of each.

2 a) Write a concise definition of neoteny and give an example.
   b) What is the possible significance of neoteny for evolutionary theory?

Read

THE COLONIZATION OF DRY LAND

Recall

1 Copy and complete table 36.3 by writing concise notes under the various headings to explain the changes necessary for animals to adapt to a terrestrial mode of life. Indicate which of these also apply to plants.

Table 36.3

	Aqueous environment	Terrestrial environment
1 Support		
2 Desiccation		
3 Respiration		
4 Locomotion		
5 Response		
6 Reproduction		

Read

EVOLUTIONARY TRENDS IN ANIMALS AND PLANTS
THE EMERGENCE OF MAN
ALTERNATIVES TO EVOLUTION

Recall

1 Using named plant and animal examples, briefly describe the main evolutionary trends.

2 a) List the characteristics of human beings that distinguish them from other primates.
  b) What are the advantages of each of these characteristics?

Review

1 Write a critical account of the heterotroph hypothesis. [20]

2 a) Distinguish clearly between spontaneous generation and chemical evolution. [6]
  b) Give an account of the hypothesis of chemical evolution and the experimental evidence which supports it. [14]

3 a) Explain why most biologists believe that life began in the oceans. [8]
  b) Give an account of the anatomical and physiological adaptations needed by an aquatic organism before it could colonise dry land. [12]

4 a) List three alternatives to evolution and make brief notes on each of them. [14]
  b) Which of these alternatives do you consider to be the most scientifically viable? Give reasons for your answers. [6]

PART 3

Techniques of biological investigation

Biology, and science in general, is a rapidly expanding field of knowledge. How do we achieve this growth in our understanding? The quest for scientific knowledge is not a haphazard process; scientists adopt a systematic approach to research. The main activities in which biologists are involved are:

- Observations
- Obtaining readings and organising results
- Forming hypotheses
- Planning and carrying out experiments
- Interpreting results and drawing conclusions

As you can see, biology is essentially a practical subject and therefore to become a competent biologist you need to learn not only what has been discovered by others, but also the methods they used. The aim of this section is to acquaint you with the methods and techniques of biological investigation.

This part of the study guide includes a series of self-assessed exercises, for some of which you will require materials and equipment. When necessary, lists of materials are provided at the beginning of each exercise. Answers for the self-assessed questions are at the back; when answers are not provided you are advised to discuss the exercise with your teacher, if possible.

Observation

Observing something is not the same as looking at it. A biological observation should be a very careful examination of some aspect of the living world. Many people have rather low powers of observation, but these can be increased by training.

Test your powers of observation by carrying out the following exercise.

Exercise 1
Examine Figure D for 20 seconds, then answer the question on the following page.

Figure D

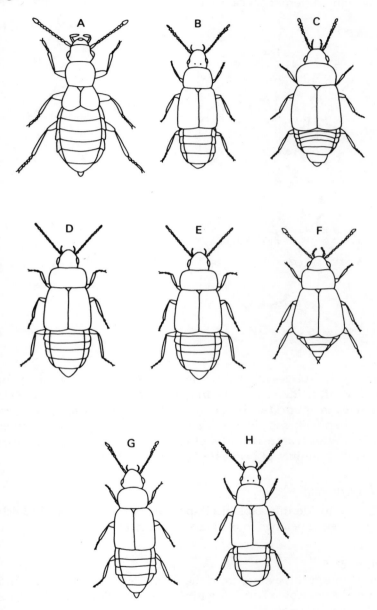

Question
How many different species of beetles are there in figure D?

Figure E

Answer
There are six species of beetle in figure D. Although there are eight drawings, **B** is identical to **H** and **D** is identical to **E**. The insects illustrated in figure D are beetles (order Coleoptera) belonging to the family Staphylinidae. Figure E shows beetles belonging to other families. What feature of the Staphylinidae distinguishes them from these other families of beetles? (Answer at bottom of page 242.)

Biological keys

One way to identify biological specimens, or to find out which family a beetle belongs to is to use a biological key.

Exercise 2
Read through the extract from the *Student's Manual* opposite and then make your own key to identify the beetles illustrated in figure E. (Answer on page 272.)

Investigation 1.1
Construction of an identification key

The diversity of organisms is prodigious and it is therefore necessary for each organism to be classified and named. They are classified according to their similarities with one another. Once a system of classification has been constructed, a method must be devised whereby other people can quickly determine the name of a particular organism. This is done by constructing an **identification key** based on the classification.

Principles involved
To illustrate how a classification and identification key can be constructed, consider the following example.

Nine students, all belonging to the same class in a school, have the following features. They are listed in alphabetical order:

Alan — dark hair, blue eyes
Ann — auburn hair, brown eyes
David — dark hair, brown eyes
Elizabeth — auburn hair, blue eyes
Jane — fair hair, hazel eyes
John — fair hair, brown eyes
Pamela — auburn hair, green eyes
Philip — fair hair, blue eyes
Susan — fair hair, blue eyes

It would be possible to classify these students in various ways. A **dichotomous classification** is one which splits them into successive **pairs** of sub-groups of approximately equal size. Such a classification is given below and from it a dichotomous key can be constructed.

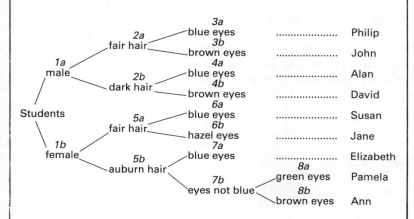

continued overleaf

> This enables a stranger, unfamiliar with the class, to quickly determine the name of any student.
>
> 1a male .... go to 2
> 1b female .... go to 5
> 2a fair hair .... go to 3
> 2b dark hair .... go to 4
> 3a blue eyes .... Philip
> 3b brown eyes .... John
> 4a blue eyes .... Alan
> 4b brown eyes .... David
> 5a fair hair .... go to 6
> 5b auburn hair .... go to 7
> 6a blue eyes .... Susan
> 6b hazel eyes .... Jane
> 7a blue eyes .... Elizabeth
> 7b eyes not blue .... go to 8
> 8a green eyes .... Pamela
> 8b brown eyes .... Ann
>
> Using the key, determine the name of the brown-eyed boy with dark hair, and the blue-eyed girl with fair hair.
>
> Exactly the same principles apply to the construction and use of an animal or plant key, as you will see if you consult a **flora**, that is a key to the plant kingdom.

Biological drawings

When appropriate, observations on the structure of organisms should be recorded by means of biological drawings. These drawings act as an aid to precise observation, and should be kept in a practical note book or file. An A4 hardback book or file with plain paper is recommended.

The extract opposite from the *Student's Manual* contains some useful hints on keeping a practical note book.

Biological drawings should be made on unlined paper with an HB pencil and should be clear, accurate records of your observations. All structures should be drawn in proportion; elaborately shaded artistic pictures are not encouraged.

Answer to question on page 240: In Staphylinids, the wing covers (elytra) are short and do not cover the whole abdomen.

Keeping a Practical Notebook

It goes without saying that the student should keep a practical notebook as a personal record of his or her investigations and observations. For this purpose a hard-backed book with alternating plain and lined paper (A4 size) is recommended.

You will be constantly faced with the question of what to enter in your notebook. The following notes are intended as a guide.

Dissection

(1) Write a *brief* summary of your technique, stressing any special points of procedure which you discovered for yourself.
(2) Draw the relevant parts of the completed dissection and label the appropriate structures. (Advice on drawing is given on p. 430.)
(3) Make sure your drawing has a heading, and indicate what view it represents, e.g. dorsal, ventral etc.

Microscopic work

(1) If you have made your own preparation (as opposed to using a prepared slide), give a brief account of your method.
(2) Draw the relevant parts of the object and label the appropriate structures.
(3) Be sure you state precisely what the object represents, e.g. whole mount, transverse section, etc.
(4) Give an indication of the scale.

Live specimens

Record your observations in the form of sketches and/or short notes, as appropriate. Get into the habit of recording your observations quickly and neatly, *while* you are observing the specimen. With experience you will learn to judge what is important and worth recording.

Experiments

Most students shudder at the thought of having to 'write up an experiment'. However, the labour is alleviated if you do as much of the writing-up as you can *during* the experiment. In biological experiments there are often odd moments when this can be done.

The format and presentation which you adopt in your notebook depends on the particular experiment. In general you should give an account of your **method**, a summary of your **results**, and a statement of your **conclusions**.

Note: The anatomical drawings in this manual are intended as a *guide* to help you identify structures which you yourself observe. You will gain very little by copying them direct into your notebook!

244 Study Guide

Figure F

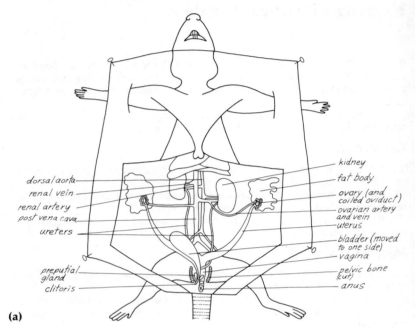

(a)

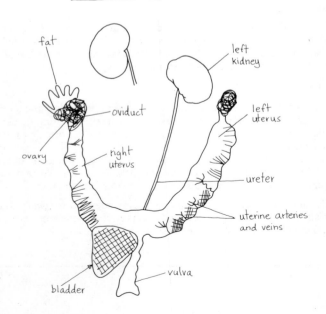

(b)

Techniques of biological investigation

Exercise 3
Figure F shows two drawings of a dissection. In an examination, one drawing received a good pass mark, whilst the other failed. Decide which is which and, using the drawings as a guide, make a list of points to be borne in mind when making biological drawings. When you have made your list, check it with the one on page 273.

Exercise 4
Materials: an apple or an onion
          a sharp knife
          unlined paper, HB pencil.
Make a drawing of a biological specimen. A suggested specimen is an apple or an onion. Cut one in two longitudinally, and make a drawing of all the structures you can see. Label your drawing using the *Student's Manual* page 263 for the apple or page 268 for the onion, or using any other biology text book for either.

Sometimes you may be asked to make an annotated drawing, which means adding short notes (annotations) to the labels. The annotations describe the functions and any other important features of the structures which have been drawn. You may need to refer to text books to acquire the information for annotations.

Figure G is an annotated drawing of a dandelion fruit (see BAFA page 405, figure 24.29).

Figure G

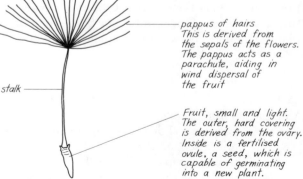

Exercise 5
Annotate your drawing of an apple or an onion.

The microscope as an aid to observation

Some biological specimens are so small that they can only be examined properly with the aid of a microscope. Such specimens may be entire organisms, such as Protozoa or Algae, or sections through tissues or organs of larger organisms. The specimens should be mounted on a glass slide and covered with a coverslip. They may be stained with dyes which show up their structures. A description of the staining methods used is given in the *Student's Manual* pages 427 to 430. The methods described cover the use of the following stains:

borax carmine, haematoxylin and eosin, safranin and light (or fast) green, safranin and light green in cellosolve.

A drawing of a whole specimen should be made either at high or low power and should include the detail observable at the appropriate magnification. Drawings of sections of specimens should follow the conventions described below.

Low power plans (LP plans) or tissue maps are drawings of sections based on what can be seen at magnifications of 10, 20 or 40. The purpose of these drawings is to map the distribution of the different tissues present in a section. It should not show the detailed structure of individual cells.

Exercise 6
Figure H shows two LP plans of a section through a dicotyledon leaf, see BAFA page 151, figure 10.21. One is an example of good LP plan, the other is a rather poor example. Decide which is which and then make a list of the points to be borne in mind when drawing LP plans. You may check your completed list with the one on page 273.

Exercise 7
Materials: a microscope
  prepared slides of a section through a dicotyledon leaf
  and a section through a thyroid gland
  unlined paper, HB pencil.
1 Use your microscope to examine the leaf section and make a fully labelled LP plan of it.

Figure H

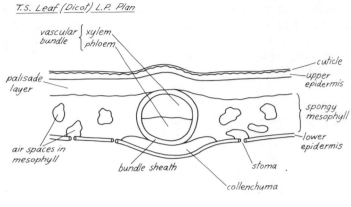

(a)

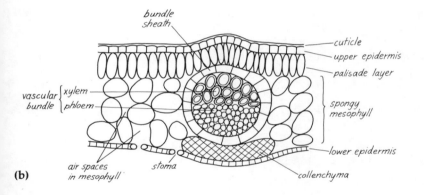

(b)

2 Examine the thyroid gland section and then make a fully labelled LP plan of two or three follicles. Use figure 18.28 (BAFA page 294) to help you interpret the section and to label your drawing. Compare your completed drawing with the one on page 273.

High power drawings (HP drawings) are made at magnifications of 100 and higher. HP drawings should be more detailed than LP plans and should show individual cells and intracellular structures such as the nucleus, chloroplasts etc.

Exercise 8

Figure I shows two high power drawings of mesophyll cells in a transverse section of a leaf, see BAFA pages 151 and 152 figures 10.21 and 10.22 A. Decide which of these drawings is the better, and then make a list of points to be borne in mind when making HP drawings of plant cells. You may check your completed list with the one on page 274.

Figure I

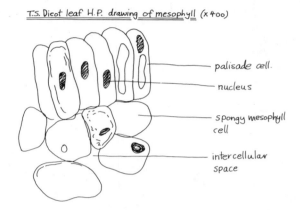

(a)

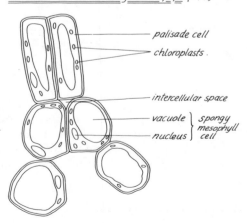

(b)

Exercise 9
Materials: a microscope
prepared slide of a section through a dicotyledon leaf and a thyroid gland
unlined paper, HB pencil.
Make HP drawings of (a) three palisade and three xylem cells, and (b) five or six cells in a thyroid follicle.

Obtaining readings, organising results

Sometimes biologists need to measure and quantify phenomena, so that their observations may be more precise and accurate. Consider for example the following observations made by two different people on the same collection of insects.

Observation one: 'The insects are much larger than any I have ever seen before.'

Observation two: 'Five insects were collected and measured. The wing span of these insects ranged from 10.00 cm to 15.21 cm, with an average of 12.60 cm.'

The second of these observations (although incomplete) is a more precise and accurate description than the first because it provides quantitative information about the insects.

The first stage in quantifying observations involves obtaining the data, which in the above example means counting and recording the numbers of insects and measuring and recording their wingspans. The second stage involves processing the recorded data, which in this case means calculating the average wingspan.

Physical quantities such as length, width, mass etc. are measured in units of the System International d'Unites (SI units). A list of SI units is given in BAFA page 660.

Exercise 10
Accurate identification of similar species often depends upon careful measurements of the sizes of various organs. Figure J (overleaf) shows the outlines of the right hind wings of several species of British dragonflies. Make and record the following measurements:
a) the overall length of the wing,
b) the width of the wing at its widest point.
It is advisable to plan how you are going to record your data before you collect it.

When you have completed the exercise, check your results with those on page 274.

Figure J

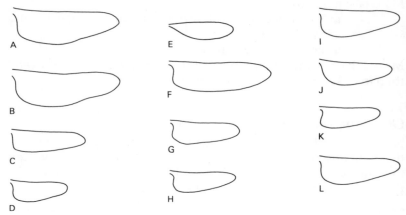

Exercise 11
 a) Make a note of the shortest and longest wing lengths. The difference between these two readings is the **range** of the wing lengths.
 b) Add together all the wing length readings and divide by the total number of species. This gives you the **mean** or **average** wing length.
 c) **Ratio** expresses the relationship between one dimension and another. The ratio between wing length and wing width can be calculated by dividing the length by the width. Calculate the length : width ratio for each species and the mean of this ratio for all species. Check your answers with those given on page 275.

Exercise 12
The table A below shows the results of three sets of measurements of the mass of a block of wood. The measurements had been made on the *same* block of wood by three students. Each student had been told that he/she was weighing a *different* block each day.

Table A

Days	Student A	B	C
1	5.25 kg	5.21 kg	5.20 kg
2	5.20 kg	5.24 kg	5.20 kg
3	5.30 kg	5.26 kg	5.21 kg
4	5.25 kg	5.30 kg	5.20 kg
5	5.10 kg	5.35 kg	5.19 kg

1 How do you account for these readings?
2 How can you minimise human error when making measurements?
3 What precautions should you take when making measurements? Bearing in mind the answers to these questions, carry out the next exercise.

Exercise 13
How long does it take for your pulse rate to return to normal after exercise?
Materials: stopwatch or a watch with a second hand
    graph paper, ruler, pencil.

Figure K *How to find your pulse*

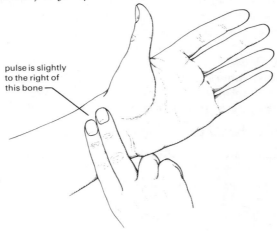

pulse is slightly to the right of this bone

Find your pulse as shown in figure K or in the *Student's Manual* page 158. Sit down comfortably and then count your pulse for a 15 second interval. From this calculate your normal pulse rate per minute. Now exercise for 3 or 4 minutes by running up and down stairs or doing pressups for example. Record your pulse rate at one minute intervals until it is back to normal.

Organise your readings (pulse beat over a 15 second interval) and your results (normal pulse rate per minute) into a table. How could you improve the accuracy of your investigations?

Exercise 14
Organise the data below into a suitable table:
Mass of fruit produced by two varieties of strawberry plant.
Measurements were made of the amount of fruit produced by two varieties of strawberry plant, A and B. Each reading is the mass (in kilograms) produced by one plant.

Variety A:
2.2,2.1,3.6,4.9,3.7,3.2,3.9,4.7,4.4,2.0,2.1,3.5,3.3,3.5,2.4,3.9,2.9,2.8.

Variety B:
4.8,5.1,4.7,4.9,5.7,4.0,4.2,4.8,4.8,4.7,5.5,5.0,4.8,5.3,4.9,5.6,5.6,5.5.

Compare your table with the one on page 276.
What is (a) the range of fruit mass produced by each variety,
(b) the mean fruit mass produced by each variety?

Graphs and histograms
Sometimes it is useful to present data in a pictorial form, as this could show up any relationships that might exist. Graphs and histograms are two useful ways of doing this.

Exercise 15: Graphs
Table B shows some readings and results of an investigation into the heart beat of *Daphnia* (the water flea). If the biological variable (heart beat) and the physical variable (temperature) are plotted on a graph

Table B *Heart beat in Daphnia over a range of temperatures.*

Temperature °C	Readings	Beats per minute (mean of three counts)
10	140, 147, 145	144
20	222, 220, 212	218
30	304, 303, 296	301
40	0, 0, 0	0

the relationship between them is immediately obvious. Figure L is an example of a graph showing these results plotted correctly, whilst figure M shows the same results plotted on a graph which shows some fundamental errors. Examine each of these graphs carefully and then make a list of points to be borne in mind when plotting graphs. Compare your completed list with that on page 276.

A similar experiment to this one is described in the *Student's Manual* page 151.

Techniques of biological investigation 253

Figure L *Graph showing the effect of temperature on the rate of Daphnia heart beat*

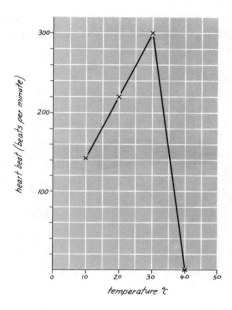

Figure M

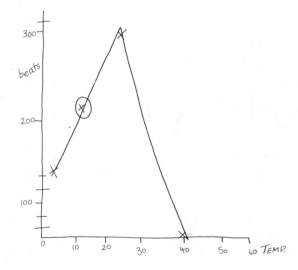

Exercise 16
Materials: graph paper, ruler, pencil.
Plot a graph of the results you obtained in exercise 15, to show how your pulse rate changes after exercise. The shape of your graph will probably be similar to figure N.

Figure N

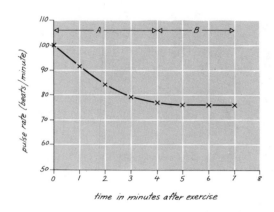

Graph to show how pulse rate changes after exercise

The first part of the graph (A) shows that the value of the variable on the y-axis (the pulse rate) decreases as the value of the variable on the x-axis (time) increases. The pulse rate is therefore said to have an inverse relationship to time. What shape would the graph be if the variable on the x-axis had a direct relationship to time?

Pulse rate depends on the passage of time, therefore pulse rate is known as the dependent variable. The passage of time does not depend on pulse rate, and so is called the independent variable.

Exercise 17
1 Examine the graphs in figure O.
Which graph shows that x
    a) is inversely related to y?, b) is directly related to y?

Figure O

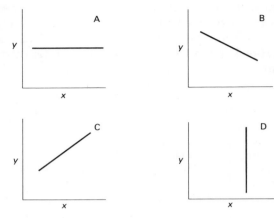

Answers are on page 277.

2 Examine the graphs in figure P and describe the relationships shown by them.

Figure P

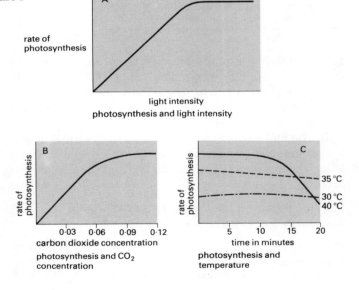

3 Table C shows some results obtained from an experiment on the digestive activity of an enzyme. The activity of the enzyme was obtained by recording the time taken for a standard quantity of enzyme to break down a known quantity of protein at a range of temperatures.

Table C *Time taken for the enzyme to break down a known quantity of protein.*

Temperature °C	Time taken in minutes (mean of three readings)	Enzyme activity
0	40	
10	11	
20	4	
30	3	
40	2	
50	6	
60	more than 60 minutes	

a) At which of these temperatures is the enzyme most active?
b) Which of the following statements best represents the relationship between enzyme activity and time:
   (i) The more active the enzyme, the less the time taken to digest the protein
   (ii) The more active the enzyme, the more the time taken to digest the protein?
c) Which of these formulae should be used to calculate enzyme activity?

   Enzyme activity α time taken to digest protein
   (i.e. activity has a direct relationship to time)

   or

   Enzyme activity α $\dfrac{1}{\text{time taken to digest protein}}$ (i.e. activity has an inverse relationship to time)

d) Calculate enzyme activity for each temperature and then plot a graph to show how enzyme activity varies with time.

Exercise 18 Histograms

Histograms are used to show the frequency of variables such as height, length, mass and so on, Figure Qa is a good histogram of the data in the frequency table on page 276. Examine figure Qa carefully and compare it with figure Qb which shows the same data, but plotted incorrectly. Make a list of points to be borne in mind when plotting histograms. Compare your list with the one on page 277 and then plot the data from the frequency table for strawberry plants, variety **B**, page 252.

Figure Q

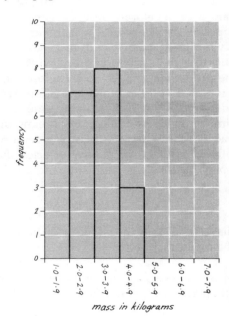

(a)

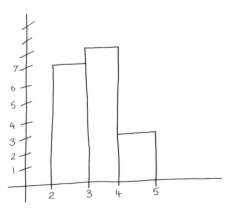

(b)

Formulating hypotheses, planning experiments, interpreting results, drawing conclusions

Observations made by biologists often prompt them to ask questions, then to make intelligent guesses (or hypotheses) to answer these questions and eventually to design and carry out experiments to test the hypotheses. For example, the observation that lung cancer occurs more frequently in smokers than in non-smokers has prompted biological researchers to ask why this should be so. An intelligent guess (or hypothesis) might be that cigarette smoking increases the chances of developing lung cancer.

To test this hypothesis, a biological researcher would plan and carry out carefully designed experiments and record all the results. The results would then be analysed to see whether or not they indicate that the hypothesis is correct. Before an hypothesis becomes widely accepted, the experimental evidence for its support must be repeated by many researchers many times.

The brief outline given above indicates that experimental biologists work in a well organised way, following certain procedures called 'the scientific method'. An outline of scientific method is given in BAFA page 3. Read through this and then answer the questions below.

Exercise 19

1 Draw a flow diagram to show the relationship between the stages of the scientific method.

2 Read through the following passage which is from a scientific journal, and then answer the questions which follow.

  a) Describe the observation that lead this biologist to formulate an hypothesis.
  b) Describe the hypothesis.
  c) Outline the experiment devised to test the hypothesis.
  d) What were the results and the conclusions from these results?

False heads fool butterfly biters

Many species use forms of protective camouflage to avoid the ravages of predators. One of the most striking examples comes from the lycaenid butterflies, which have evolved a pattern of wing markings, and twin tails which look like antennae, to give the appearance of a head at the tail end of the body. What evolutionary advantage does this provide? Many people have speculated that it may misdirect predators into attacking the less vulnerable tail end of the butterflies. Now, Robert Robbins, of the US National Museum of Natural History of Smithsonian Institution, has carried out a study which lends weight to these speculations.

Robbins has studied two sets of these butterflies, one sample of more than 1000 specimens from 125 species, observed in Colombia and the other, 400 specimens representing 75 species, in Panama. Species with the classic 'false-head wing', he says, are five times as likely to have sustained wing damage from a 'deflected predator attack' (that is, a lump bitten out of the back of the wing) as are species with a less pronounced false-head pattern on the wings (*American Naturalist*, vol 118, p. 770). The proportion of specimens Robbins saw with hind-wing damage differed from random expectations, confirming the hitherto unproven speculations of biologists about the value of the patterning, and evolution at work. □

The purple hairstreak butterfly, with its false 'eye' and 'antenna' on the wing

Exercise 20
Read through the observations below and then answer the questions that follow.
1. The cut surface of a sliced apple often turns brown when exposed to the air. The browning can be prevented either by covering the sliced apples in water or by sprinkling lemon juice over the apples.
   a) Suggest an hypothesis to explain these observations.
   b) Design a controlled experiment to test your hypothesis.

2. A group of pregnant ewes ate poisonous plants that caused damage to the brains of the lamb foetuses they were carrying. The lamb foetuses were not born at the usual lambing time.
   Birth in sheep appears to be triggered by a rise in the level of adrenal hormones in the blood of pregnant ewes.
   a) Suggest an hypothesis to explain these observations.
   b) How could you test your hypothesis?

3. It is well known that swans take in small pieces of gravel to help grind up food in their digestive system. In recent years, there has been a marked decline of the swan population in parts of the River Thames. During the same time, there has been an increase in the number of anglers in these parts of the river.
   a) Construct an hypothesis to explain these observations.
   b) How would you test your hypothesis?

4. *Chondrostereum purpureum* is a fungus which causes a fatal disease called silver leaf disease in a number of trees, including those grown for fruit production. Pear trees, carrying infections of the harmless fungus *Trichoderma* show some resistance to silver leaf disease.
   a) Suggest an hypothesis to explain this.
   b) How could you test your hypothesis?

Exercise 21
Read through BAFA page 6, and answer the questions below.
1. Explain, with reference to the experiment on *Amoeba*, the meaning of a control experiment.

2. The following is an account by an experimenter who had seen an advertisement for a plant food called Magigrow. The advertisement had promised that Magigrow would make plants grow twice as tall as normal. The experimenter conducted an investigation in order to find out if this claim was true. Below is his

account of the investigation. Read through it once, and then answer questions (i) and (ii).

'I obtained two potted plants[a] and placed them on a windowsill[b]. To test the effect of Magigrow[c], one plant was watered daily with the recommended dose of Magigrow.[d] I left the second plant unwatered[e]. At the end of several days[f] the first plant had grown by 10 cm[g] the second (untreated) plant had not grown[h] at all. Therefore, Magigrow does make plants grow'[i].

(i) Comment on whether you think the conclusion about Magigrow is correct.

(ii) How could the experiment be improved?

Now read through the account a second time and then answer questions (iii) and (iv). Answers on page 277.

(iii) Note the underlined parts of the account. The questions below refer to each of these parts (a) to (i).
a) How could the description of the plants used be improved?
b) Describe how you could improve the experimental conditions by making them similar and controlled.
c) Explain why this is not an adequate description of the purpose of this investigation.
d) How could you improve this account of the watering procedure?
e) Explain why this is not an adequate control experiment.
f) How could you make this statement more precise?
g) How could this statement be made more precise?
h) What evidence would be required to make this into a scientifically acceptable statement?
i) Is this a justifiable claim? Give reasons for your answer.

(iv) Explain the difference between an hypothesis and a theory.

3 Experiments are usually written up under these headings: title, aims, materials, methods, results, discussion/conclusions. Below is a student's account of an experiment. The experimental procedure and conclusions show a number of errors. Read through the account critically, and then answer the questions that follow.

'I noticed that earthworms appear to feed on plant materials, such as dead leaves. I decided to find out how they are able to do this. An earthworm gut was dissected out and washed through with water to obtain gut enzymes. This mixture was then tested to see if it was capable of digesting starch. This was done by adding the enzyme mixture to starch solution. At five minute intervals, a drop of the enzyme/starch mixture was removed and tested with iodine to see if the starch was being digested. After twenty minutes, the mixture gave a negative test for starch, indicating that earthworms have starch-digesting enzymes in their guts.'

- a) Describe (i) the observation, and (ii) the hypothesis in this investigation.
- b) Re-write the account under the five headings suggested above. Add any extra experimental details you think are necessary.

You may find it helpful to discuss this exercise with your teacher.

How to tackle examination questions

There are, broadly speaking, two kinds of questions in A-level biology theory examinations; structured questions involving fairly short, specific answers and essay questions which are more general.

A. Structured questions

Study these three examples of structured questions:

1   a) Name the phylum to which each of the following belong:

   a leech ....   a frog ....

   a jelly fish ....   an amoeba ....
   (4 marks)

   b) State *three* reasons why an earthworm and a ragworm are classified in the same phylum.

   (i) ....

   (ii) ....

   (iii) ....
   (3 marks)

c) Give *three* ways in which an earthworm differs from a ragworm.

(i) ....

(ii) ....

(iii) ....
(3 marks)

2 The volume of air entering and leaving the human thorax can be measured by a spirometer and recorded with a kymograph. The figure below represents part of such a kymograph tracing.

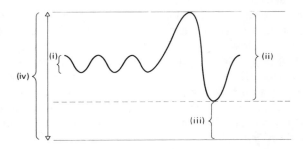

In each case briefly explain what is represented by:

(i) ....

(ii) ....

(iii) ....

(iv) ....

3 a) List the functions of lignified tissues in plants.   (4 marks)

Give an illustrated account of
b) the types of lignified tissues to be found in a *named* herbaceous angiosperm,   (10 marks)
c) the distribution of these tissues in this plant.   (6 marks)

Questions 1 and 2 are probably the most straightforward. The questions are clear and direct and the answers require only a few words. Answering this type of question should not take very long. In many examinations consisting of this type of question, candidates have no choice of question, and should attempt them all. Before putting pen to paper, it is useful to scan the whole paper, but not more than about five minutes should be spent on this. You will probably find there are some questions you can answer straight away, others that you are not sure about and some that you have not much idea about. Do the ones you are confident about first, and you may well find that the answers to the more difficult questions come to you as you are answering the easy ones.

This type of paper also often includes questions in which you have to plot a graph from given data. If you bear in mind the advice given about graphs on page 276, you should be able to score maximum marks for this type of question.

Question 3 represents a type of structured question which is rather less straightforward, and this is reflected in the amount of time you should devote to the question. The instructions usually advise that you should spend approximately 30 minutes on your answer. It is essential when answering questions of this type (and even more so when tackling essay-type questions) to spend a few minutes in calm, careful planning before beginning to write the answer. At the start of the examination, read the instructions carefully, and then read through the whole paper before starting to answer any of the questions. In some examinations you may be expected to make a choice of question; make sure you are clear about this before you start.

Again you will find there are some questions that you are confident about, others you are not so happy about and some you can eliminate straight away. Make your selection carefully; you may find it helps to start with your second best question. This should help you avoid the major pitfall of 'getting carried away' and spending too long on a question about which you know a lot. (This is also true for essay questions.) Whatever happens, do not spend significantly more than the suggested time on each question and always answer the required number of questions.

Opposite are two typical students' answers to question 3 (a) and (b). Read through each one and then answer the questions which follow.

Answer 1

The functions of lignified tissues are mechanical and vascular.

The main types of lignified tissues in plants are xylem tissues and sclerenchyma.

Xylem tissues contain two types of lignified cells, which are vessels and tracheids

A xylem vessel in T.S.

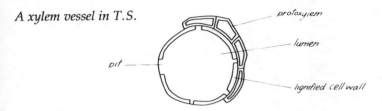

Tracheids are similar to xylem vessels, but are not cylindrical in section.

Sclerenchyma cells have evenly lignified walls.

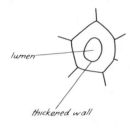

Answer 2

a)

Lignified tissue	Function
Xylem	Vascular tissues, transporting water and mineral salts.
Sclerenchyma	Mechanical tissues providing support.

b) Lignified tissues consist of cells in which the secondary cell wall is impregnated with lignin. Lignin is a complex carbohydrate, commonly known as wood, and it strengthens cell walls and makes them waterproof. Examples of lignified tissues found in *Helianthus* are xylem tissues and sclerenchyma.

Xylem tissues are made up of vessels and tracheids. The vessels develop from procambium into elongated cells which join end to end. The end walls between cells eventually break down and the lumen of each cell interconnects with that of each cell immediately above and below it, see the diagram below. At the same time, lignin is laid down over the primary cell wall. Lignin can be laid down in a number of different ways.

Diagram to show lignification of xylem vessels

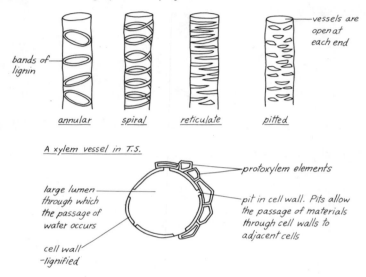

Tracheids are similar to xylem vessels but are not open at each end and are not cylindrical in section.

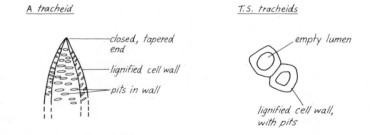

Exercise 22
Copy table D into your note book. Complete the table by writing in the information given in answer 1 and answer 2.

Table D	Answer 1	Answer 2
Functions of lignified tissue		
Names of lignified tissues		
Lignin		
Named angiosperm		
Description of lignified tissues Xylem		
Sclerenchyma		

What is the main difference between these two answers?

The main difference between the two answers is that answer 2 contains more detail. One reason why some A-level students do not do well in examinations is not because they do not know any biology (although obviously this is sometimes the case), but because they are not aware of the differences between an O-level and an A-level answer. In answer 2 it is assumed that the reader has little knowledge of biology and therefore all biological terms are explained in some detail. In answer 1, the student has assumed that terms such as mechanical, vascular and lignified are understood by the reader, and has not explained them.

How do you go about answering a question like question 3? First, it is a good idea to analyse the question and identify the key words. The key words for part (a) are underlined below:

List the functions of lignified tissues in plants.

For (b) and (c) the key words are:

Give an illustrated account of
(b) the types of lignified tissues to be found in a named angiosperm,
(c) the distribution of these tissues in this plant.

In order to answer this question, you need to know what lignified tissues are, to be able to make diagrams of them and be familiar with a suitable named angiosperm. You will also have noticed how the marks are allocated; 4 for (a), 10 for (b) and 6 for (c). On this basis you should not spend more than 5 minutes on (a), 15 minutes on (b) and 10 minutes on (c). When asked to list information, do just that (as in answer 1 but not answer 2).

Remember to indicate which part of the question you are answering. See for example, in answer 1, the student has not written in (a) or (b) on the answer, and it is difficult to know where (a) ends and (b) begins. It is obviously important to know this because of the way the marks are allocated.

If asked to use annotated diagrams, do not waste time by giving a written description of the structures you have drawn (as in answer 1). Annotated diagrams should be used instead of written descriptions, not as well as.

Exercise 23
Bearing all these points in mind, answer question 3(c) on page 263.

Exercise 24
Below is another example of an A-level question and an answer to part (a) of the question.
Read through both, and then answer the questions that follow.

Question
Using a named example for each, describe how the male and female gametes are brought together in
(a) a mammal,   (10 marks)
(b) an angiosperm.   (10 marks)

Answer
In mammals, the males produce the sperm and the females produce the eggs. Fertilization is internal and this is brought about by an intromittent organ called the penis. The production of gametes is controlled by hormones in mammals. Female mammals produce eggs only during the oestrous cycle and this is the only time that reproduction is possible. Prior to fertilization, copulation must occur. The penis becomes stiff and erect and the spermatozoa, together with seminal fluid, are ejaculated into the vagina.

The spermatozoa swim up the vagina, into the uterus and eventually reach the oviduct. Eggs are liberated from the ovary into the oviduct and it is here that fertilization occurs. Usually only one sperm unites with the egg and it is then impermeable to other sperm.

1 Identify the key words in the question.
2 Analyse the answer to see how these key words have been tackled.
3 Rewrite the answer, including more detail to bring it up to A-level standard.

B. Essay questions

Below are some typical essay-type questions based on those found in A-level examinations.

1 Discuss the roles of carbohydrates in the lives of organisms.
2 'Water is essential for life.' Discuss.
3 Explain the function and importance of photoreceptors in the lives of mammals.

The instructions on most papers advise that you should spend between 35 and 45 minutes (including planning time) on these questions.

The most obvious feature of essay questions is that they are *unstructured*. They have not been broken up into subsections with the mark allocation for each indicated. All you know is that you should spend about 35–45 minutes on your answer and that the total mark allocation will be (usually) either 20 or 25 marks.

The points made previously about reading and selecting the questions carefully are particularly relevant to this type of question, and thorough planning of your answer is of great importance.

The structure of your answer, the order in which you present the main points, the diagrams you use, and the amount of detail you go into are all decisions you have to make *before* you start writing your answer, and for questions like this you should reckon to spend about five minutes planning your essay.

In the planning stage, you should jot down, in very abbreviated note form, the main items of information to be included in your answer. You can do this in linear or pattern note form, or in the form of a table. In deciding what information to include, imagine you are aiming your essay at an intelligent non-biologist. Do *not* assume a specialist knowledge of biological terms and concepts on the part of the examiner. When you have noted all the essential points, you should then decide in what order they should be presented, and what diagrams, if any, you are going to include.

The steps in the process of planning an answer to question 1 would be as follows:

1 Analyse the question

Key words: Discuss, roles, carbohydrates, organisms
Discuss here means 'Give a full and reasoned account of . . .'
Roles of carbohydrates, concentrate on **functions**, rather than **structure**, of biologically important carbohydrates.
Organisms, present a balanced account, dealing adequately with *both* animals and plants.
(N.B. Up to now, the planning process has been mainly mental – it is not suggested that you would write all this down, except perhaps to jot down the key words.)

2 Plan your essay

A question such as this, involving discussion of plants and animals, could be planned in tabular form:

Table E *Roles of carbohydrates*

Plants	Animals
Manufactured in photosynthesis from CO_2 and H_2O 1 <u>Source of Energy</u> Hexoses e.g. Glucose main respiratory substrate	Cannot be manufactured – major item in diet
2 <u>Storage</u> i.e. energy store – polysaccharides Starch – storage organs e.g. tubers, roots etc. Seeds. Sucrose – sugar cane (stems), sugar beet (roots). (disaccharide)	Glycogen – liver, muscles. (mention structural difference starch/glycogen)
3 <u>Structural</u> E.g. Cellulose (long unbranched chains of glucose form microfibrils) <u>Lignification</u> 5C sugars form part of nucleotides	Chitin contains glucosamine

Diagrams: none really necessary.
(You will find information about carbohydrates in BAFA page 64.)

Once you have written down the main points in this way, you can begin to see the essay take shape. When you are sure you have not left anything out, and have decided on the structure of the essay, you are ready to start writing.

Exercise 25
Now make your own answer plans for questions 2 and 3. You may compare your plans with the ones given on pages 278–9, but do not assume these are the only 'correct' versions – yours may be just as good.

Answers to questions

Exercise 2

Key to identify the beetles in figure E

1. a) Head not visible from above, completely covered by thorax..B.
   b) Head clearly visible, not covered by thorax......................2.
2. a) Antennae fine and filamentous, segments not clearly visible..C.
   b) Antennae thick, segments clearly visible........................3.
3. a) Antennae very long, as long as head, thorax and abdomen combined...E.
   b) Antennae not as long as body .... 4.
4. a) Head with beak-like snout (rostrum)............................A.
   b) Head without rostrum..5.
5. a) Points of jaws (mandibles) crossed................................F.
   b) Points of mandibles not crossed...................................D.

Notes

This is not the only correct key to the beetles in figure E. There are many other features that could be used to form the basis of an equally good key, such as the shape of the various parts of the body, the presence or absence of hairs or spines on the legs, the size and shape of the eyes, and so on. The important thing when constructing a key is to use features which can be easily observed and do not vary between individuals of the same species. Size and colour differences, for example, are only useful when there is no variation within the species. Similarly keys based on information about the biology of the species, such as the kind of plant it feeds on, are not very useful, as this information may not be available when you are trying to identify a dead specimen.

Exercise 3
Figure Fa is correctly drawn.

Points to be borne in mind
1. Outlines should be drawn with one neat, even, unbroken line.
2. Drawings should be accurate (note the inaccuracy of the blood vessels on Fb).
3. Shading should be kept to a minimum. Where used, it must be neat.
4. Drawings of parts of organisms should be placed in context, where possible. In this case, an outline of the entire specimen shows the relative position of the urinogenital system.
5. Label as many structures as possible using straight lines which originate on the structure.
6. Label lines look neater if they are either drawn parallel to each other, or radiating from a central point.
7. Label lines should not cross each other.
8. The drawing should have a title, the organism's name and classification should be given where appropriate and the scale of the drawing should be given.

Exercise 6 – Low power plans
Figure Ha is a correctly drawn LP plan. In addition to the points considered in exercise three, the following should be borne in mind when drawing LP plans:
1. The drawing should show, in outline, the shape of the specimen.
2. Tissue regions should be indicated by means of simple outlines only; the LP plan should not show individual cells.

Exercise 7

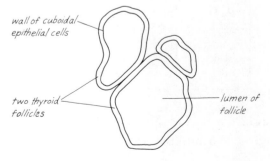

Your LP plan should be an accurate map of two or three thyroid follicles. It should bear no resemblance to figure 18.28A in BAFA. This figure is a *diagram*, which is a simplified and stylised representation of thyroid gland structure. Diagrams can be used to illustrate essays, reports etc., but they should *not* be used as a record of practical work.

Exercise 8 – HP drawings

Figure Ib is drawn correctly.
In addition to the points considered in exercise 3, the following should be borne in mind.
1. Generally, only a small number of cells (3–6) need be drawn for each tissue.
2. HP drawings should show the position of neighbouring cells accurately, they should not show cells appearing to interlock in an unnatural manner.
3. Plant cells have cellulose cell walls. These should be indicated by a double line. Animal cells do not have a cell wall, and the cell membrane should be indicated by a single line.

Exercise 10

Table F

Species	Wing length (mm)	Wing width (mm)
A	28.0	9.0
B	28.5	9.5
C	19.5	6.0
D	15.0	5.5
E	17.0	5.0
F	27.0	8.0
G	18.5	6.5
H	17.5	5.5
I	21.5	7.5
J	19.5	6.5
K	16.0	6.0
L	21.5	7.5

Exercise 11

a) Range = 28.5 mm (B) – 15.0 mm (D) = 13.5 mm

b) Mean wing length = 20.73 mm
Did you spot the catch here?
I and L are actually the same species. Therefore the mean wing length is the total of A – K divided by 11, the true number of species. If you failed to notice that I and L were identical, and counted them as separate species, you will have obtained a slightly different answer. This also applies to your answer to (c).

c)

Table G

Species	Wing length : wing width ratio
A	3.11
B	3.0
C	3.25
D	2.73
E	3.40
F	3.37
G	2.85
H	3.18
I	2.86
J	3.0
K	2.66
L	2.86

Mean Length : width ratio = 3.03

Notes
1 A ratio is an indication of how one measurement relates to another, i.e. how much larger (or smaller) the length is when compared to the width. Therefore ratios are expressed as a simple number without units.
2 When the ratios were calculated, they ran to several decimal places. It is quite acceptable to round them up or down to the nearest single decimal place.

Exercise 14
The data should have been organised as shown in table H, overleaf, which is a frequency table.

Table H *Frequency table of mass of plants of variety* **A**

Plant A Class	Frequency
2.0–2.9	7
3.0–3.9	8
4.0–4.9	3

The large number of individual measurements of mass have been organised into smaller groups or classes, e.g. 2.0–2.9, 3.0–3.9 etc. The number of times (or frequency) a plant falls into each of these classes has then been recorded in the table. Frequency tables are useful for recording and organising large numbers of measurements taken from individuals. If you did not draw a frequency table, do so for the measurements of variety **B**.

Exercise 15 – Graphs

1. Always use graph paper.
2. Give the graph a title.
3. Always draw the axes at right angles to each other. The horizontal axis is the x-axis; the vertical axis is the y-axis.
4. Generally, the biological data is plotted on the y-axis, the physical or environmental data is plotted on the x-axis.
5. Where appropriate, the origin should be at the point where $x = 0$ and $y = 0$. Sometimes it is not advisable to do this, for example if the readings on the x-axis are of a different magnitude to those on the y-axis, or if x or y never becomes 0.
6. Indicate the scale divisions on both axes and the value of the intervals.
7. Select a suitable scale so that your plotted data extends over most of the graph.
8. Label both axes fully using only SI units and accepted abbreviations.
9. Plot points clearly so there is no doubt as to which point you have plotted.
10. Connect points with a line which gets as close to as many points as possible. This is called the best straight line.
11. If the points on the graph represent a curve, join them up by a smooth curved line, drawn as close to as many points as possible. A curve suggests a relationship between points; a jagged line between points suggests that each point is an isolated event.
12. Distinguish clearly between different sets of data using different symbols or lines.

Exercise 17
1. a) B
   b) C

Exercise 18 – Histograms
1. Bear in mind points 1–8 given for exercise 15.
2. Make the blocks the same width. The area of the block represents the frequency of the variable. If blocks are the same width, the heights of the blocks are proportional to the areas and to the frequencies.

Exercise 21
iii a) The description of the plants should provide enough precise information about them to enable another researcher to repeat the experiment. For example, the species (and possibly variety) of plant should be identified, and their age and information about the conditions under which they were grown prior to the experiment should be given. Relevant details here would be the size of pot, type of compost, temperature and daylength conditions.
   b) Conditions on a windowsill are unlikely to be identical for both plants, and would probably fluctuate considerably. You could not be certain that differences in the growth of the plants was *not* due to these differences and fluctuations. You should try to ensure that conditions for both plants are as identical as possible. This could be done by placing the plants in a growth chamber, room or greenhouse in which environmental conditions are controlled.
   c) The hypothesis the experiment is supposed to test is that Magigrow causes plants to grow to twice their normal height. The investigator should have been more precise about this aim.
   d) Again, more precise details should be given: 'One plant was supplied, at the same time each day, with the manufacturer's recommended dose of Magigrow, i.e. 2.5 g Magigrow dissolved in 1 litre of tap water. 50 ml of this solution were poured onto the surface of the compost in the pot.'
   e) The experiment is intended to investigate the effect of Magigrow, not water, on the plant's growth. An appropriate control experiment would be to water the second plant with an equal volume of tap water without Magigrow added.
   f) The precise number of days should have been given.

g) Some indication of how growth was measured should have been given, e.g. 'The height of the main stem, measured from soil level to the tip of the apical bud, increased by 10 cm.'

h) Even if the length of the main stem had not increased, this is not proof that *growth* had not occurred. Growth could have occurred in side branches, leaves, or roots, for example. A better way of measuring overall growth would be to record changes in the mass of the plant. The data gathered could be tabulated and presented as a graph. Alternatively, the investigator could have said 'The second (untreated) plant showed no increase in the length of the main stem'.

i) This is a very large claim to make on the basis of a single experiment involving one plant. It would really only be justified if consistent results were obtained from a large number of plants, using adequate controls. Even then it would only be valid for the particular species of plant used.

Note

You may have noticed the inconsistency in the experimenter's use of verbs in the account, e.g. 'I obtained two plants . . .' and 'one plant was watered daily . . .' It is generally considered that the second, more impersonal style is more appropriate for accounts of experiments, i.e. 'Two potted plants were placed . . .' 'The second plant was left unwatered . . .' and so on.

Answer plans for the essay questions on page 271 (Exercise 25)

2 'Water is essential for Life.' Discuss.

i.e. main topic: importance of water to living organisms:

1 Constituent of protoplasm – c. 90% H_2O.
2 Medium for metabolism – all biochemical processes take place in an aqueous medium.
3 As a reactant in biochemical processes e.g. photosynthesis, hydrolysis.
4 As a medium of transport: Of gases, nutrients, secretory and excretory products, cells, heat etc.
5 As a medium for the dissemination of reproductive bodies, e.g. gametes (especially sperms) larvae, embryos, cysts, spores, fruits, seeds etc.
6 Support (a) External – buoyancy (b) Internal – Turgor (especially important in plants but also in many animals e.g. hydrostatic skeleton of annelids, eyeball and penis in mammals).

7 As a source of food, both in itself $H_2O + CO_2 \longrightarrow C_6H_{12}O_6$ and as a source of dissolved minerals for plants and animals. Also special case of *filter feeders*.
8 As a source of O_2 – directly for aquatic organisms, terrestrial spp. must provide moist *respiratory surface* for gas exchange.
9 Protective function – fluid filled coelom acts as shock absorber for protection of internal organs. Also high heat capacity protects against overheating.
10 Lubricative function – coelomic fluid lubricates movements of internal organs (e.g. between pleura in thoracic cavity). synovial capsules, mucus, tears (also protective) etc.

Concluding paragraph: Life on earth believed to have evolved in aqueous medium, protoplasm is aqueous, life existed only in aqueous environment for millions of years. Very difficult to imagine life forms evolving in the absence of water.

Diagrams: none necessary.

3 Explain the function and importance of photoreceptors in the lives of mammals.

Key words: Explain, function, importance, eyes, mammals.
Plan
Function
1 Structure of mammalian eye:
Diagram. Horizontal section – *Annotate* functions of parts
Brief account of accommodation, structure (diagram if time) and function of retina.
Importance
Very important receptors in lives of most mammals (other important sense-smell) few exceptions (e.g. bats, moles).
Importance of sight:
1 To locate food – herbivores discriminate different plant types
– carnivores locate and catch prey.
2 To avoid and escape from predators and other dangers.
3 To identify and communicate with members of same species – mates, offspring, competing individuals, e.g. warning signals, courtship displays etc.
4 Role of photoreceptor in regulating behaviour – e.g. night/day, seasonal changes in daylength etc.